CW01216911

# The Healing Nature of
# WATER

First published in Great Britain in 2000
by Gaia, an imprint of
Octopus Publishing Group Ltd
Carmelite House
50 Victoria Embankment
London EC4Y 0DZ
www.octopusbooks.co.uk

An Hachette UK Company
www.hachette.co.uk

The authorized representative in the
EEA is Hachette Ireland,
8 Castlecourt Centre, Castleknock Road,
Castleknock, Dublin 15,
D15 YF6A, Ireland
(email: info@hbgi.ie)

This edition published in 2025

Text copyright © Octopus Publishing
Group Ltd 2000, 2005, 2025

Distributed in the US by Hachette
Book Group
1290 Avenue of the Americas, 4th and
5th Floors, New York, NY 10104

Distributed in Canada by Canadian
Manda Group
664 Annette St., Toronto, Ontario,
Canada M6S 2C8

All rights reserved. No part of this work
may be reproduced or utilized in any
form or by any means, electronic or
mechanical, including photocopying,
recording or by any information storage
and retrieval system, without the prior
written permission of the publisher.

Charlie Ryrie asserts the moral right to
be identified as the author of this work.

ISBN 978-1-8567-5579-5
eISBN 978-1-8567-5580-1

A CIP catalogue record for this book is
available from the British Library.

Printed and bound in Great Britain.

10 9 8 7 6 5 4 3 2 1

Publisher: Lucy Pessell
Senior Editor: Tim Leng
Designer: Isobel Platt
Assistant Editor: Samina Rahman
Production Controller: Sarah Parry
Cover Illustration: iStock, CPD-Lab

This FSC® label means that materials
used for the product have been
responsibly sourced.

### Acknowledgements
Huge thanks must go to Jane Waters for
her enthusiasm, her knowledge and her
support. Thanks to all the people who
have helped me gather material for this
book and given time to help make it
better, particularly Patrice Bouchardon,
Jennifer Greene, Alan Hall, Sandra
Gibbons, Julian Jones, Merrily Lovell, Sir
Peter Guy Manners and John Wilkes.

### Disclaimer/Publisher's note
The techniques, ideas, and suggestions
in this book are to be used at the reader's
sole discretion and risk.

Always observe the cautions given, and
consult a doctor if you are in doubt about
a medical condition.

# The Healing Nature of
# WATER

**CHARLIE RYRIE**

Gaia

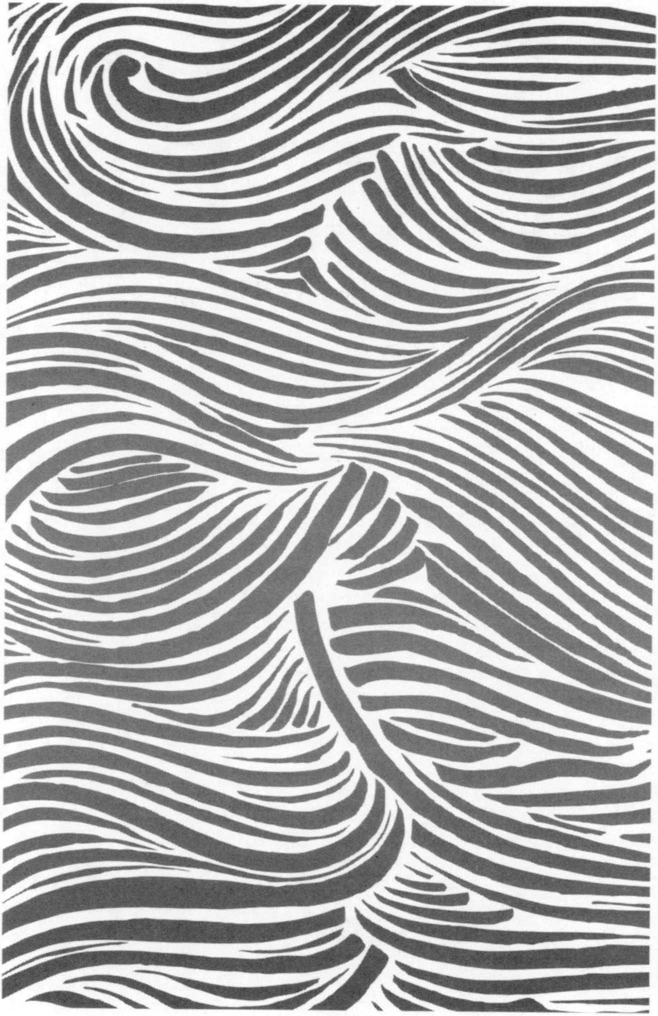

# Contents

| | |
|---|---|
| Introduction | 9 |
| Chapter One: Sacred Water | 20 |
| Chapter Two: Water for Life | 39 |
| Chapter Three: Using Water's Vital Energies | 78 |
| Chapter Four: Bathing | 112 |

# Foreword

By David Lorimer
Director
The Scientific and Medical Network

The Seven Lakes region of the Rila Mountains in Bulgaria is associated with Orpheus, the mystical source of spiritual thinking in Ancient Greek culture. When one is there, one can almost hear him play; one has the strong sense of being in a sacred landscape permeated by sublime presences watching over the pilgrims. The water there is of two kinds, each of which speaks its own symbolic language: the spring and the lakes. The spring is a key symbol in the writings of the Bulgarian sage Beinsa Douno (Peter Deunov, 1864–1944). In the summer camps of the 1930s one of the main tasks was the cleansing and laying out of springs, using the local quartz to construct sites to draw water and drink. The most arresting of these places still features a sculpted pair of marble hands, through which the crystalline water flows. Nearby is an inscription enjoining the traveller to give freely like the source.

The source, however, can only give if it receives: the cycle can only be sustained if the receiving or taking is in equilibrium with the giving. This is the great lesson of the spring as of modern ecology. If the mountain peaks symbolize masculine striving, the lakes represent not only feminine receiving and holding, but also, as the names of two lakes suggest, purity and contemplation. Their transparent stillness seeps into the soul with deep refreshment and renewal.

To read this book is not only to be reminded of the sacred significance of water and its importance for our health, but is to have one's eyes opened to water's multiple dimensions and the extraordinary extent to which we in the developed world complacently take it for granted. Every student of Ancient Greek philosophy remembers the quotation from Thales, the founder of the Pre-Socratic movement, that water is the material cause of all things. Also, the insight from Heraclitus that one cannot step twice into the same river.

Great civilizations and cities have been built on water and have in some instances been destroyed by its excess or shortage. These civilizations recognized their dependence on water and incorporated it in various ways in their myths. The modern danger, especially in the Middle East, is of water wars breaking out on account of chronic shortages, thus accentuating our ultimate dependence on water – think of the strategic importance of the Jordan and the Nile. One message of this wide-ranging book is that we have lost our connection not just generally with the Earth, but especially with water, and that we urgently need to regain it.

In a scientific and medical sense, this means that we tend to consider water in a mechanistic and quantitative fashion. We pay attention to water supply and its treatment with a variety of chemicals, but we are not concerned with its vitality or its fundamental role in maintaining good health. The author explains the work of Viktor Schauberger and Johann Grander on the importance of spiral movement in running water as a conductor of vitality. Vitality? Surely this is just a form of vitalism, long since discarded by a biological tradition that cannot even define life itself! We recognize the dangers of stagnant water but fail to appreciate the importance of live water for our health and wellbeing. Precious little research is pursued in this field, but I believe that we are now seeing the emergence of a new, qualitative approach to science that will begin asking a different set of questions.

Water is associated with both life and death: the Gospels speak of the living waters of truth and baptism by water is a central Christian rite. If water brings life, it can also bring death, and the crossing of a river is often associated with the passage from this life to the next. Many near-death experiencers report that the crossing of a river symbolizes the point of no return. They remain on the bank, like Hermann Hesse's *Siddartha*, who learns to listen from the river, who realizes that the river is everywhere at once, at the mouth and the source, and who eventually surrenders himself to the scream, 'belonging to the unity of all things.'

To read this book is to become aware of water and its significance in a new way, to open one's eyes to its many aspects and to appreciate it as never before. The author has done us a great service in reminding us of many things we already knew but had never fully integrated into a complete picture.

# Introduction

## The Lifeblood of Gaia

The crucial thing that distinguishes Earth from other planets is water. The Earth, or Gaia, was created from it and is sustained through it. Water is everywhere: about 70 per cent of our planet is covered by water. We were born from it, and all life depends on it. We are all made of around 75 per cent water, and rely on water for life. We are watery creatures living on a watery planet. Water is the essence of life.

Our life and health, and the life and health of Gaia, are entirely dependent on water. It carries all present life, the seeds of future life and the memory of past life in its flow. It is the mediator between life and death, between being and nothingness, between health and sickness. But somewhere over the years we have lost our connection with this life-giving and life-supporting substance. Familiarity is partly to blame. When the first pictures of Gaia were sent back from space we were stunned by the incredible beauty of our planet, a shining blue sphere, bathed in a delicate swirling membrane. We could see the blue water of the oceans and the white vapour of the clouds endlessly circulating around and over the brown land masses, the whole suspended in an endless inky ocean. The image was a beacon of light in the darkness of space, something to treasure. Now such photographs are commonplace, and the image of Gaia is reduced in many people's minds to little more than a corporate logo. It has lost its magic.

For most people on Earth, water also has lost its magic. Because it is so universal we take it for granted; we have stopped thinking about it, let alone appreciating it. The water available to so many people throughout the world at the turn of a tap is far removed from the crystal clear bubbling water of a natural spring, the reflective water of a mountain pool or the deep water of an ancient well. Our modern demands and processes have effectively neutered it and removed its essential character. Instead of revering its once-sacred nature, we have gradually allowed water to become degraded and dirtied. Exploited, manipulated and polluted, this fountain of life has become one of the great channels for transmitting illness – and so we 'clean' it with chemicals and recycle it without first revitalizing it.

Newborn humans are 97 per cent water | Adults are about 75 per cent water | The human brain is 75 per cent water | Bones are 22 per cent water | Tooth enamel is 2 per cent water

The present state of water is something that we should all think about. Our relationship with the water that we consume is two-way. We are all made of water and we must respect water in order for it to support us. If handled incorrectly water can itself become diseased, and then it imparts that disease to all other organisms. So when we put mistreated water into our bodies, our metabolism becomes unnatural, chemically and energetically unbalanced. Our biochemical systems become disturbed, the messages to our brains confused. When our body is out of balance our emotions are constantly reacting according to 'fight or flight', stressing us.

Most of us are aware of the problems with polluted water supplies, problems due to environmental degradation through our own actions or inactions. Many people now buy bottled water to drink, or filter tap water, not trusting the quality of the water piped to us – yet the quality of bottled waters is also questionable. We know that it is unsafe to bathe at some beaches; we probably also know that parts of the world are already well on the way to dramatic water shortages owing to careless or unnecessary overuse of water in domestic and industrial processes.

We may feel powerless about some, or all, of these problems. Yet, if we take time to look at the marvellous substance that is water, we soon realize that water is itself the source of power, and if we appreciate its qualities we can become empowered.

Water is a colourless liquid that enables us to see the colours of the rainbow. It has no form but gives form to everything. It is billions of years old but can constantly rejuvenate itself. It is at the beginning and end of every cycle of life and history. It provides a constant interchange between the Earth and all the other planets in the cosmos. Every drop is a microcosm of the universe, carrying information from ancient eras and from worlds that we cannot yet understand. It is a liquid, a solid and a vapour. It is everywhere and in everything. It holds the key to our health, and we hold the key to the health of water.

The natural world is balanced, each part of it in harmony with the other parts, positive balanced by negative, yin with yang, masculine with feminine, the visible world reflecting the invisible. The amount of water in our bodies reflects the amount of water on the planet. This suggests how human beings are made in an image of the wider world – we are tiny microcosms within the macrocosm.

# Gaia's Water

It is commonly believed that life began when a supernova exploded in the Milky Way and atoms dispersed to form a new planet, Earth. This was so hot that nothing could settle on it and a dense cloud of methane, ammonia and carbon dioxide covered it for millions of years. In time this poisonous early envelope boiled off into space and as everything cooled a new atmosphere developed, dominated by hydrogen and carbon dioxide. These atoms eventually rearranged themselves and joined together to form water vapour and carbon. As the planet continued to cool this turned to rain, and for millions of years the rains came down. Water crept into each crack, bored its way through different surfaces, filled up every empty space, and the process of life began – all species evolving from carbon and water.

The Earth is bathed in over 1,337 million cubic kilometres (320 million cubic miles) of water, circulating around the planet as water, vapour, rain, snow or hail, creating and sustaining life. Liquid water is always moving, in streams, rivers, lakes and oceans, above and below ground, or borrowed by plants and animals, then turned to vapour, forming great circulatory streams moving over the Earth.

As vapour, clouds are continually transporting water around the world until they fall as rain to replenish the Earth's liquid store. At any one time, roughly 50 per cent of Earth is covered in some type of cloud as 1,420 cubic kilometres (340 cubic miles) of water is evaporated daily from land and ocean surfaces. And every day, the same quantity falls back to earth. The story of water is a cyclical tale of rhythm and movement, a tale unfolding since the dawn of Earth's time. As well as surface water, there is so much additional subterranean water that it could cover the whole surface of the Earth to a depth of 305 metres (1,000 feet). Despite this, only 1 per cent of the Earth's water is available for daily use: 97 per cent is in the sea; 2 per cent is deep frozen, locked in icebergs and glaciers. In its solid state – ice, frost, snow, glaciers and icebergs – water can be locked up for an hour, a season, or for as long as a billion years. Of the 1 per cent of the Earth's water, about 98 per cent is used for industrial purposes and only about 2 per cent for the nutrition and health of every living thing. Because we depend on water for life, we must take care of what we have.

We communicate with water on different levels – physically when we see a river, an ocean or a pool, or when we are thirsty or dirty – but it also speaks to us on a deep subconscious or spiritual level. We are all attracted to water in some way. We don't often recognize the connection, but it is very significant. Water is not only the foundation of life, but also the link between our planet Gaia and other worlds. Our nearest star, the sun, and our satellite, the moon, work together to create the ebb and flow of tides, and the Earth's gravity pulls water toward the oceans. Water is always on the move, each drop responding to the pull of the moon and sun. Even the water in a teacup has its own infinitesimal tides, responding to the call of the planets. The blood in our veins, the water in our cells, every fluid in our body responds to the same cosmic forces. Water links us to the rest of the universe.

> During one summer day a single willow tree can use and lose over 2,272 litres (500 gallons) of water. Some 25,000 cubic kilometres (6,000 cubic miles) of water moves each year through the plants covering the Earth's surface; this is almost the same as the water carried to the sea by all the rivers of the world.

# Our Watery Origins

It seems that water has been a part of our emerging consciousness since human life began. Some people believe this relationship with water stems from the possible aquatic origins of the human species. Early evidence of humans was discovered in footprints in a once watery area in the East African Rift valleys. About ten million years ago the waters gradually receded forcing the aquatic dwellers to take to the ground to survive.

We share many of our defining traits with sea mammals. We have, for example, large brains; we are naked and experience orgasm; we communicate primarily through the voice; and our body fat is evenly distributed in a subcutaneous layer rather than stored in humps, pouches or the like. This distribution of fat, it has been argued, only makes sense for a creature that swims regularly and needs to be buoyant but reasonably streamlined. An early diet of seafood could account for humans' large brains relative to other primates.

But perhaps the most often quoted evidence for our aquatic past comes from the fact that babies are perfectly happy in water for the first year of their lives after birth. They behave calmly, with no apparent fear of drowning – they never attempt to breathe under water, and simply paddle calmly up to the surface when they go under. Of course, some would say that babies are simply reliving their life in the womb.

Whatever the truth of such theories, there can be no doubt that the way of life of most peoples in the world has led to a loss of consciousness of water. We have cut ourselves off from its symbolic message as well as destroying its physical potential.

# Ancient Wisdom, Future Health

In the cycles of nature wind turns to rain, rain to earth, earth to sea and back to clouds again. The constant element through all these changes is water. Centuries ago people lived surrounded by nature, never questioning the health-giving properties of flowing streams, invigorating waterfalls or welcome rains. They accepted and, over time, came to venerate and treasure the powers of water. Most modern cultures are now cut off from direct communication with the elements. Instinctively, though, our senses and subconscious still recognize – and yearn for – links with a natural environment.

Some of the stress we suffer derives from a rebellion against the complex weight of unnatural living. Many cultures have become divorced from the natural world, from spiritual realities, and from anything that cannot be understood through our immediate senses. We have learned to be sceptical of things that cannot be 'proven' through rational scientific methods.

> The story of expulsion from the Garden of Paradise, symbolized throughout history and mythology as the expulsion from a beautiful watery place to a barren fiery land, may reflect a dim memory of human migration from a wetland 'paradise' and watery lifestyle to drier and less congenial conditions.

Traditional science tends to approach things from a linear and reductionist perspective – not the best tool for discovering the truth about water, which is a fundamentally holistic substance. Each phase of water's cycle reflects the other phases; no one phase can be deemed more important than another; water is vital to every cycle in life. It is part of a whole, and always seeks to return to a whole. There is nothing linear about water. When astronauts first looked down at our watery planet from space, they saw a sphere covered in an ever-changing pattern of water vapour. If you magnify any single drop of water you will see those same shifting movements at work inside each sphere. We can't understand a flowing stream by looking at all its different facets in isolation – the physics, chemistry, biology and geology – because each affects the others and are interrelated. Understanding water by division is like putting together an image without an idea of the whole picture.

Earlier cultures seem to have understood water much better than we do today. Pure water, for example, has always been treasured: ancient Chinese saved the water from glaciers in jade vases; Incas and Aztecs put water in jars of obsidian; witch doctors in Africa used quartz crystal. Modern science now recognizes that the minerals in these containers affected the water because high concentrates of silica allow water to maintain its structure and prevent it from becoming polluted or weakened. But these peoples didn't have to analyze the water to know about these properties; they knew that putting water into certain containers kept it healthy and pure. They knew that water, constantly on the move, was affected by, and affected everything around it. They didn't need scientific tools to recognize that water collects, stores and transfers physical and vibrational information or energy as it moves.

If you throw a stick into a pool of still water, if a fly alights on its surface, a fish pokes its head through or a duck dives, the surface of the water is momentarily shattered. A series of ripples move outward, markedly near the centre, becoming less defined as they spread out, until they vanish. The ripples are vibrations, resonating at a different frequency from the bulk of the water, transporting messages from the point of disturbance to the wider whole. They are channels of energy, changing the energy of the water around them.

This is obvious when you see it happening, yet nowadays we are claiming that water's role as a channel for vibrational information is a new rather than an ancient discovery. We are beginning to see life as a result of vibrations associated with the phenomena of resonance and magnetic fields. Water is the channel that allows the transmission of vital information in every cellular and extracellular exchange. In the next century water could again become one of the best cure-alls. But to rediscover its healing potential we have to try to understand it.

Water was at the root of classical medical philosophy in ancient Egypt, India, China and Assyria, as well as Greece, all of which linked the individual with the cosmos, seeing the two as interrelated, each reflecting the other. In the Chinese tradition water is the key to the five elements: wood is created from water, which encourages the growth of plant matter; fire is encouraged by wood; earth comes out of fire; metal comes from earth; and water arises out of metal (condensation). Water was seen to link the five levels of human existence: physical, vital energy, emotional, mental and spiritual. Water does affect us physically, mentally and emotionally; it is the vital fluid that allows vital energies to flow freely. Without healthy water, we cannot be healthy.

# The Highest Element

The ancient Greeks revered water as the medium and possibility for all elemental change, the highest of the four elements from which the world was made. Water was crucial in systems of magic and alchemy, where the two principal elements were fire and water. Fire was not perceived as a tangible flame, but as something more subtle, such as electricity. Water's dense physical form is represented by rivers, streams and oceans; but in alchemy the subtlest form of water is magnetism. We are just starting to re-acknowledge the importance of electricity and magnetism for health, something that earlier systems of knowledge just knew. We are only just beginning to learn – or relearn – about water's role in transmitting electromagnetism.

Our disconnection from the world around us has been a gradual process over centuries. An increasing devotion to certain types of rationality made us suspicious of the spiritual or religious realm, and distrustful of anything that could not be proven by the application of linear scientific laws. But rationality and spirituality need not conflict – we don't have to cast off what we know in order to understand or appreciate the world. The Japanese, for example, probably the most technologically advanced nation in the world, have retained a deep connection with water through thousands of years, sensing no conflict between scientific rationale and the spiritual.

The cultural heritage in the West accentuates the division between the material and the spiritual worlds. Yet, even in Northern Europe, tens of thousands of rational people flock to shrines such as Lourdes in France each year, believing in the healing powers of its waters. Rituals associated with water are practised all over the world as a matter of course. Deep down we all want to retrieve or maintain a balance between the spiritual and the material parts of our lives, to tap in to something that is ultimately beyond our rational understanding of the world.

For those of us who are still bound by a need for hard scientific proof, one of the most exciting developments at the end of the 20th century was modern science's new ability to explain and help us understand the potent effects of ancient practices. Science is now beginning to recognize levels of more subtle energies such as the electromagnetic fields that surround us, the energy frequencies that course through our bodies, the vibrations that affect our life processes. Just as the past lives of water affect its present life, past interpretations of water's powers often offer us keys to present understanding.

Water provides a mirror of society, it reflects everything that we put into it. At the moment the water on Gaia too often reflects a sick society, polluted and degraded through our actions. But we can change this reflection. This book is an invitation to recapture the magic of water, and life. If we appreciate water and treat it well, health will follow.

Water is much more than a concrete phenomenon with specific scientifically described properties and abilities. It is a mysterious living creature that nurtures all life. Water is a powerful carrier, mediator and even producer of energy. It has the ability to link, transform and capture physical elements and subtle energy forces, and it operates on a physical and a spiritual level.

# Chapter One
# Sacred Water

If we take the time to stop and think about water, we realize that it is a magical substance with a special meaning for everyone. Sadly, few of us do stop and think, and the gulf between ancient and modern ways of looking at water has grown wider and wider over the centuries. Some cultures, such as the Cherokee Indians, continue to revere the Earth in the way they always have done. They see water as the Earth's lifeblood, the rivers and streams that course down mountains and hills and through valleys are the arteries and veins that allow blood to course through the body. Water is part of their consciousness. On the other hand, ask a modern Western child what they understand water to be, and they will probably say it is something we drink that comes out of a tap. An educated adult might describe water in terms of its physical properties, as a colourless liquid, or as a 'scientific anomaly'. Of course, these descriptions are true, but they leave out so much.

If we change the question slightly, and ask people how they feel about water, we can begin to approach the way water affects us on so many levels. Most of us love to relax in a bath when we are tired, or feeling out of sorts or anxious. We take it for granted that a bath will make us feel better. A bathroom is often seen as a sanctuary where we can retreat from the world, ignore outside pressures and just enjoy the sensations of bathing our body in water, feeling the ripples on our skin, hearing the splashing noise of our slightest movements in the water. We don't think about it, we just know that a bath can be a healing experience. We don't need to analyze our feelings, we just know.

# Connecting With Water

We know that walking by the sea, or by a rushing stream, or past pools with fountains makes us feel good. We know water is the most refreshing drink when we are thirsty; we know to drink a glass of water if we have been drinking alcohol, or overindulging; we know it feels good to stand under gentle rain and feel the drops on our skin. And we know children are fascinated by water; they want to play with it, and in it. They love the way it moves, the way it feels, the way it runs into spaces and fills them up; they are intrigued by its very wetness. When most people think of relaxing or taking time out from work, water comes into the picture somewhere – it could be a trip to the seaside, swimming in a pool, walking by a lake, pond or river, staying on an island, boating, water sports, dolphin watching, fishing. Venice is one of the most visited cities in the world, the Niagara Falls are one of the world's greatest natural spectacles. Would people cross continents to see the Niagara Falls if they were just a tract of sand in the same place?

Everyone wants to be connected to water in some way, yet we don't recognize the fundamental attraction, nor question it. Over centuries we have become curiously separated from our deepest feelings about the world around us, compartmentalizing the different facets of our life into physical, practical, emotional and religious boxes.

# Bringing Back the Sacred

A common cry in today's fast-moving society is the plea to bring the sacred back into everyday life in order to reconnect with ourselves and with the world as a whole. We are aware that there is something missing in our lives, but we don't know how to reclaim it. Once there was a general understanding, or even an unspoken assumption, about what was sacred or spiritual. Now we all have different ideas of what the terms mean. Some people equate the spiritual with the religious; others, particularly Westerners, see it as pious expression about religious events or rituals, or as reference to a specific way of behaving, something outside everyday experience. But notions of piety and sacredness are not shared the world over. Some cultures realize we don't have to cut ourselves off from modern life to experience the sacred – for example, it is not unusual to see a Hindu taking a holy bath at a sacred site while scrolling on their smartphone!

The everyday and commonplace can also be sacred, everything is worth reverence – this idea has always been central to many belief systems originating in the East, and it is no coincidence that many Eastern societies have a far more respectful attitude to water than that shown in the West. From similar roots Eastern and Western societies developed along markedly different paths over the last two millennia, although there has always been an interchange of ideas. While the West pursued economic dominance and power for many centuries, the second half of the 20th century saw renewed interest in Eastern beliefs and practices, and a renewed search for spirituality.

People have worshipped the Ganges for thousands of years. Just as the Nile dominated ancient Egyptian religion, so the Ganges, flowing from the Himalayas to the Bay of Bengal across the north Indian plain, has been a focus for devotion since prehistoric times. Following the birth of Hinduism, numerous myths emerged to account for the origin of the great river. Hindus believe that after 60,000 sons of King Sagara were punished for their arrogance and burned by Vishnu, the divine goddess Ganga was brought down from the heavens to purify the ashes. She performed the ceremony at the delta of the Bay of Bengal by creating a great river to cleanse the Earth and free their souls, and the river was named after her. Ever since, Hindus have paid homage to the Ganges, believing that all sin is cleansed by the ritual bath in the river, and those that drown in the river are believed to be reborn among the gods.

> Agrarian societies have a variety of practices to catch and hold the attention of the gods and keep the water flowing. Rainmaking rituals are part of the culture of most tribal agricultural societies. Ngoni and Ronga tribes in Africa sing ribald songs to the water gods. In the Caucasus, girls yoke themselves to a plough and drag it to a dry riverbed. In Transylvania, they sit naked on a harrow. In parts of southern India, women have to catch a live frog and tie it to a fan.

When we talk of the sacredness of water, we refer not only to specific beliefs about water, or particular sacred sites. We also try to encourage people to view water as the magical, mysterious and all-powerful substance that gives us life. We each know axioms to the effect that what we put into something, we get back; if we treat water with respect and reverence, it will repay us generously. Western society underwent radical changes in the 16th and 17th centuries, the beginning of what is known as the Age of Rationality, culminating in demands for the separation of religion and science. From then on empirical rationality became the aim, to be rid of every trace of myth. Before that time even Western people understood the world in a different way. They didn't separate the elements but knew that the heavens and Earth were connected; they knew it was important to treat the Earth well; they accepted that religion and spiritual beliefs were an integral part of everyday life. They respected both the deities and the elements. Beliefs didn't rely on empirical evidence, but on practice and deep intuitive understanding. Mysteries were an accepted part of life.

If human beings deny the mysteries of life, they are not beginning to live to their true potential. It is possible that much of the stress in society today – human and environmental – comes from separating rationality and spirituality, of what is empirically provable and what is mysterious. Many people's lives continue to become more complex and fast moving, as they try to balance the demands of careers, families and self-development, while being bombarded with aspirations and role models and surrounded with pressures. No wonder increasing numbers of people are taking the waters or are searching for waters that heal. We all need water in order to live and breathe, but we also need it on an emotional and spiritual level. Everyone can live in a dialogue with water that links physical, emotional and mysterious worlds.

# Sacred Beginnings

Water exists at the beginning, and returns at the end, of every cosmic or historic cycle: it has the ability to confer life, uphold life and transform death into new life. Throughout human history water has been connected with mystery, or used as a symbol of mystery. It has always been a symbol of life.

Water and life were once inseparable. Every ancient culture revered water; all the earliest deities were water gods. In Sumerian the word *mar* meant sea, but it was also the word for womb; *a* was the word for water, and also meant sperm, conception and generation. The Hebraic language includes the ideogram *Mem*, deciphered as mother, life, womb or sea. 'Everything was water' say Hindu texts, and in Tantric manuscripts water is *prana*, the vital breath that brings life. Water and life were one and the same.

Water precedes all forms and upholds all creation, and every society has stories accepting that the world came out of water, describing this event with all kinds of watery creation myths: Sumerian, Arkadian and Babylonian creation stories describe how the Earth was born out of the union of Apsu and Tiamat. Apsu represented the primordial sweetwater chaos that surrounded the stars in a wide stream, and Tiamat the bitter-tasting salty sea. The Chaldeans and Persians worshipped the goddess Anahita, who personified Earth. She was believed to have been created from seminal fluid flowing from the point where the watery stream surrounding the Earth met the stars.

> Rivers and sacred waters have always attracted sacrifices and offerings, and these practices still continue in many agricultural societies where water is the obvious difference between life and death, prosperity and destruction. The Masai of West Africa throw a handful of grass into the water each time they cross a river; the Buganda of Central Africa traditionally offer coffee beans.

Hindu mythology has many variations of watery creation myths. In one version Narayana was floating peacefully on a cosmic sea creature on the cosmic waters, and the Tree of Life sprang from his navel, the centre of his being. Another tradition replaces the Tree of Life with a lotus, out of the centre of which Brahma was born. And another tale describes Vishnu lying on a cosmic serpent floating on the primeval waters, and willing the Earth into existence. In another version Vishnu dives down to the depth of the waters and draws the Earth up from the abyss.

Similar myths appear in European folklore and several North American Indian tribes believe the creator sent animals down to the bottom of the sea to bring up the mud, out of which he created the Earth. Plains Indians believe that all was still until the Old Man appeared floating on a raft and willed the Earth into existence out of the water. The Pima Indians of New Mexico believe that Mother Earth was impregnated by a drop of water that fell from a cloud.

One Japanese story describes how a giant carp swimming in the unbounded sea thrashed his tail so wildly that the Earth was born out of the shock waves. Chinese creation stories describe how water represents the primordial chaos, the totality of possibilities of what life could be.

The Greeks had vast numbers of water deities and associated myths, where water was love and power, and symbolized birth, beauty and destruction. One myth describes the birth of Aphrodite, the goddess of love and beauty. Her father Uranus (heaven) and her mother Gaia (Earth) together symbolized the great stream believed to gird the Earth. When Uranus mistreated Gaia, she asked her son Cronos to overthrow his father. So Cronos castrated Uranus and scattered his seed upon the ocean. This floated on the waves and became the sea foam that gave birth to Aphrodite.

# Taking Water for Granted

The connection between fertility and water is obvious, but modern industrialized people, living far from the land and the seasons, often forget the link. When it rains we may complain of the inconvenience, or how it makes travelling difficult, or interferes with our plans. Yet, when lack of rain causes the authorities to impose limitations on water usage, the loudest complaints are from people who can no longer water their gardens – the link between water and fertility only recognized when it impinges on our own private demands.

The irony is that, because we have taken water for granted for so long, we are facing the prospect, even in the industrialized West, of consistent shortages of water of sufficient quality to maintain our health or the health of the environment. We have forgotten how to respect water. Until we regain an attitude of reverence, we are threatening our life and the life of the planet on which we live.

# Water and Transformation

All water carries within it the possibility of death, as well as the possibilities of life. Just as water begins its life cycle underground and then returns to the ground after it falls as rain, so water transforms death to allow life to begin again.

If you live near a river, you respect it. You will know how the landscape changes and life contracts when the river dries out too much. You will know from bitter experience how a storm can change your life by flooding your house, or by destroying your property and the habitat for other living creatures. You will also know that water has a life of its own because you are familiar with the different facets of its character. You will recognize all too well the power of water, and its ability to communicate.

Civil engineers have developed ways of minimizing the harmful impacts of water's rages – flood defences, irrigation schemes and pipelines. We don't turn to water for our transport nearly as much as we used to, so our daily needs do not rely directly on natural water-courses. We don't depend on our rivers in the same obvious ways as our ancestors did. But imagine if the rivers were the prime source of life, health and fertility, of communication and transport, as well as containing the potential to take life away through storms and floods. Then we would treasure them, fully recognizing their independent life.

When people live close to nature, they depend on the elements, and water is respected for what it is, life-giving and life-threatening, a force for good and evil. Major rivers have been sacred in many societies throughout history. The Nile lay at the heart of early Egyptian religion, and sacred to Egyptian society; it was later venerated by Greeks and Romans. In India many rivers are still regarded as sacred, but the Ganges is the most sacred of all.

Floodwater has often been seen as especially sacred, as a symbol of cleansing the old and returning to a pre-formed state, ready for rebirth. Many ancient cultures believed that the greatest tribute they could give to a deity was to be drowned in a flood or a river, consequently being united with the river god.

# Crossing the Waters

Water has always been associated with death as well as life, as the transforming element, and many rivers only exist in mythology as waters that have to be crossed on the way to the afterworld. Buddhists and Taoists embrace the idea of crossing the great waters on the journey to Nirvana, and in parts of West Africa people believe that the dead are carried over three rivers which separate this world from the next – senior members of the Yoruba tribe are buried in canoes to prepare them for their journey.

Some societies still recognize water as a powerful symbol in death; the process of laying out and cleaning a dead person's body can be a ritualistic washing away of the sins, or may symbolize the purification of the soul before it enters the next world. Many North American Indian tribes place a bowl of water near the dying person so that their spirit may leave the temporal body and enter the water, from where it can be reborn. They directly recognize the universality of water and the need to reunite with the watery element.

Burial in the Ganges is still a vital part of Hindu culture, and devout Hindus worldwide send the bodies or ashes of their loved ones across continents to be buried in the sacred river. Funeral pyres burn day and night on the banks of the Ganges at Varanasi, prior to the ashes being scattered on the river. The souls of those whose ashes are placed in the river are believed to be freed from the sins accumulated in their previous lives. Pilgrims also come in their thousands to worship and immerse themselves in the holy river.

## Baptism

Holy water throughout history is healing water, understood to heal symbolically as well as, in some cases, physically. Baptism is possibly the universal symbol of purification and regeneration.

Baptism was first introduced in rites and ceremonies of the Great White Brotherhood in Egypt. An avatar (the equivalent of a priest, a person who points a member of their religious community in the direction of god) had learned through meditation that water would purify humans spiritually and physically. The ritual is also rooted in Shinto, Confucian and Hindu customs of bathing in sacred waters symbolically to cleanse all sin. It was the baptism of Jesus by John that prepared him to accept his mission on Earth as the Son of God, as if the magic contact with water conferred his role upon him.

In modern Western culture, baptism may be the only ritual associated with water that still has an obvious spiritual element. It symbolizes rebirth and belonging, the idea of access to unknown but desired mysteries, and a promise of entry to an afterlife and hence immortality.

People may not step inside a church for any religious reason yet they often want their children to be baptized or christened by a priest, even though it seems rather unconnected with other areas of their life. This may seem a parody of an ancient ritual. But it's as if the parents continue to believe deep down that the blessing of the water will confer on their child some kind of protection against evil forces. There are calls to separate baptism from the church – one English priest has given up his calling in despair after being asked more than once to perform baptisms in bathtubs, so that people wouldn't be inconvenienced by having to attend church! Why do we cling on to these customs with such tenacity? There seems to be something in all of us that needs to recognize our spiritual nature, a desperate desire to find some meaning on a deeper level. A ritual such as baptism is so strong that people still want to be involved in it, even though its symbolism, on a conscious level, may have become weakened so far as to become all but lost.

# Sacred Waters

People have held on to the idea of sacred water, despite cultural and idealistic changes in many societies. The search for – and discovery of – healing waters has never ceased, and communities have grown up around springs and wells. Water cults have been universal, throughout the world, throughout history.

Water came to be known as beneficial or destructive, and had to be revered. The spirits of wells were usually seen as benevolent, while those of running water were more tempestuous. Almost identical beliefs appear at different times in primitive Polynesian, tribal African and developed Western societies – that you must not look into running water because you look into God's eye, or because you look into your soul, and to dream of your reflection is to dream of death.

Numerous myths exist of seductive sirens, ambiguous mermaids, undines, nixies and silkies, all beautiful female creatures waiting to captivate and drown unwary youths and steal their mortality. Springs, wells, pools and fountains were usually occupied by more helpful beings, often blessed with the gift of prophecy.

The Roman philosopher Seneca stated:

> *'Where a spring rises or a water flows, there ought we to build altars and offer sacrifices.'*

In Celtic Europe water cults were particularly common, but the Christian church clamped down on the practices, believing the power that people invested in the water deities detracted from the message of the Christian religion. So water cults were strongly suppressed, but the lure of sacred water was still acknowledged as churches or religious communities were often built around springs and wells, with the church appropriating those sacred waters for itself!

Water cults have never died out, although they have often been absorbed into the general culture and religion of an area. People still make pilgrimages to wells for healing emotional and physical problems. Lourdes is visited by an astonishing six million people a year! One of the most sacred Islamic sites is a spring in the centre of Mecca, a central part of the Hajj pilgrimage. According to Muslim legend Abraham's son Ishmael and Ishmael's mother Hagar ran out of water while they were crossing the desert; Hagar climbed several mountains in search of water, while Ishmael sifted sand through his fingers. Suddenly, life-giving waters bubbled out of his hand, and became Zamzam, the holy spring at the heart of Mecca.

Sacrifice was often a part of the earliest veneration of springs and wells. Animals were offered to appease the water gods and spirits. The practice of leaving offerings is still common at sacred wells where people take the water to ask for cures. Common offerings range from a piece of clothing from the person seeking a cure to religious statuettes or gifts of food for the water spirits. In some parts of Ireland, a land with hundreds of small springs and wells, it is not unusual to come unexpectedly across pieces of ribbon tied to plants and discover a tiny spring, known locally for some specific healing powers. At some of the larger sacred water shrines, special rooms have been constructed to contain the offerings. Offerings reinforce the fact that human beings are in dialogue with water. It is a two-way process. What we give to water, it will give back to us.

Wells were often renowned for their peculiar characteristics and people would throw special things into them. These wishing wells attracted coins, pins, buttons, pieces of broken pottery, etc. One well, renowned for curing toothache, needed hazel twigs or nuts to work its powers. Another, thought to have the ability to convert everything into precious metal, required pine cones. White stones were offered to wells that could influence the weather.

Giving thanks to the spirit of a well for its pure clear water was a widespread ritual, particularly in England. Every summer, people decorated, or 'dressed', their local well with flowers and other votive offerings. Often, the flowers were arranged in elaborate designs and images. The custom is still practiced in Tissington, Derbyshire.

# From Ancient to Modern

The Japanese have revered water throughout their history, largely because of the persistence of Shinto attitudes, combining religion, myths and folk beliefs with moral injunctions to maintain harmonious social and political relations. Insistence on ritual and actual purity are central to the Shinto ethos. The spirit most Japanese bring to their relationship with water incorporates a connection with the values of their ancestors, although they may not necessarily think of bathing as a spiritual act, nor consider their passion for hot springs an expression of their religious beliefs.

Blessed with thousands of bubbling hot springs because of their islands' volcanic geology, the Japanese see these springs as gifts from the gods to the Earth – some of their oldest sanctuaries are built close to these sacred springs, which have been revered for thousands of years. We now know that the different springs, or *onsen*, do have very specific healing qualities according to their different mineral content, so ancient practices really do help bathers to feel better in body and mind.

In Japan, open-air baths combine the warmth of a natural spring with beautiful scenery for spiritual refreshment. These *rotenburo*, which literally means 'a bath amid the dew under an open sky', are popular all year round – including the heart of winter when they are surrounded by snowdrifts.

Just as ancient beliefs can influence modern practices, science and the sacred can work in the same direction. New research is indicating the way that water encapsulates and transmits energy, and scientists have even proved that sacred or healing water is both qualitatively and quantitatively different from other waters.

Two Italian scientists spent months measuring the electrical fields' of water at different healing wells in Europe, including Lourdes, and discovered significantly different results from those at 'ordinary' springs or wells. They deduced that this was not only because healing springs are often rich in minerals such as germanium (which helps to maintain high oxygen levels in water, and therefore significant energy). But it may also be partly the two-way communication between the water and those seeking cures: if you project loving thoughts into water, it affects the water.

# Bringing the Sacred Home

In Persian and Arabic lore, paradise was an enclosed garden containing flourishing fruit trees, beautiful flowers and flowing water. Outside the garden walls lay a burned landscape, a sandy hell. Without water, there could be no paradise. Water was the divine gift, symbolizing heavenly possibilities. Paradise gardens in Islamic culture always include gently flowing streams, rippling pools or fountains.

No Japanese garden is complete without flowing water, often in the form of a small fountain or bubbling pool, beside which people can stroll or stop and meditate. Water has always had a special spiritual role in meditation and relaxation. It is almost impossible to feel depressed or angry in the presence of a flowing stream, a splashing fountain, a breaking wave or a roaring waterfall. Water will touch all of us, if we approach it right.

Unfortunately, few of us spend enough time relaxing, let alone meditating or trying to get in touch with our spiritual nature. We all know how we can be uplifted or inspired by listening to a particular piece of music, or by looking at a painting, a sculpture, a piece of architecture that moves us. We may feel elevated after reading a particularly moving book. We use the word 'moving' freely. It implies that an experience moves us away from day-to-day reality on to another level of experience. Some experiences are genuinely uplifting, but others encourage us to retreat into a deep dark place of fear or depression.

Water moves through every level of existence – it falls from the heavens to earth and reaches deep into the subterranean world below before returning to the world above. To fulfil our potential as human beings we also need to operate on different levels – we all need insights and inspiration from a heavenly realm, awareness from the earthly realm, and confrontation and resolution of fears from a subterranean level. If we treat water with reverence, we will recognize that it has many powers – to inspire, to heal, to give and take away life. It acts on our physical body and our emotions, on a conscious and a subconscious level. We need to treasure water so that it can nourish us.

> *The sage's way, Tao, is the way of water. There must be water for life to be, and it can flow wherever, and water, being true to being water, is true to Tao.*
> *Those in the way of Tao, like water, need to accept where they find themselves, and that may often be where water goes to the lowest places, and that is right.*
> *Like a lake the heart must be calm and quiet with great depth beneath it.*
> *The sage rules with compassion and his word needs to be trusted.*
> *The sage needs to know how to flow around blocks and how to know a way round like water and how to find the way through like water.*
> *Like water the sage must wait for the moment to arise and be right.*
> *Water never fights; it flows around without harm*
> (from Chapter Eight, Tao Te Ching)

If we realize how much we rely on water, we will be more careful with it. Unfortunately it retains only a vestigial sacred nature for most people, and in most places. Our exploitation of water has been continuing for centuries, but has become most obvious in the last 70 years. Now there is piped water on tap for much of the world, and piped sewerage systems. We don't need to think about how we get clean water, how we dispose of it once we have dirtied it, or how much of it we use. Water can't possibly express its true nature if we don't take care of it.

This doesn't mean that everyone needs to drastically change their lifestyle straight away. We can't all suddenly change our living spaces or our jobs, for example, but we can become more conscious of the world around us and reconnect with a more spiritual side of life. And we can start this process by being more conscious of water and reconnecting with its sacred nature.

It is not difficult to bring a little bit of paradise into our own homes and workspaces. You can incorporate a small water feature in your garden or yard, or even indoors. But be careful where you place it. To a practitioner of the Chinese art of *Feng Shui*, which literally means 'wind and water', water is a very powerful force because it absorbs and stores energy. The location of water in your home and garden can present difficulties because water is such a powerful feature. For example, a water feature in the garden to the front of your house is much more beneficial than one at the rear.

You can take time to reflect or meditate by water – a pool, a stream, a lake – and there are a number of ways to ensure that the water we use is healthy (see Chapters Two and Three) and able to fulfil its potential. Water is alive – it has reflected and carried the beliefs of every society since time began. It works in mysterious ways, but, without it, all life stops. Once we begin to understand its true nature, we will realize that it is the key to spiritual, emotional and physical wellbeing. If we reconnect with water, we reconnect with the forces of life.

# Chapter Two
# Water for Life

## What is Water?

Water is so much a part of life that we tend to ignore it and look elsewhere for the magic ingredient that will increase energy, health and vigour, so making us live longer. But the key to health and longevity is remarkably simple: it is water. When we come to understand water fully we will have the key to all life's processes. Water is not a self-contained and isolated substance; its characteristics depend on its situation; its structure enables it to react with other molecules; and, as we shall see, to retain imprints of its previous experiences.

In some areas of the world, well documented by anthropologists, people seem to age slowly, remaining healthy until well over 100 years old. Degenerative diseases are almost unheard of. One area is the land of the Hunza people, in the Karakoram Mountains north of Pakistan; another is a mountainous area of Ecuador. Hunza water is ice-melt from ancient glaciers, water that has been locked up for millions of years. It is probably the purest form of water found on Earth.

Hunzas drink pure water all their life, consequently they live a long and healthy existence. Hunza water doesn't contain any of the usual mineral salts found in mountain springs or well water as it hasn't travelled through the hydrological cycle and picked up minerals and other information along the way. Most water contains mineral salts dissolved in the water, but in Hunza water minute mineral clusters, including a high proportion of carbons and silicates, are suspended in the water, forming an ideal structure to support and enhance life.

Water has many secrets it is only just beginning to share with us again. Over recent centuries understanding of water has focused on a scientific viewpoint. Yet the answers lie in a more integrated vision, taking nature as the inspiration and model. To appreciate its resilience and adaptability we need to understand its form and energy, and the way every droplet is a microcosm of the universe, reflecting and encompassing nature.

Every child learns at school that water is a chemical compound of two simple and abundant elements – a dipole molecule consisting of two positively charged atoms of hydrogen (H) and one negatively charged atom of oxygen (O). The formula $H_2O$ may be simple, yet the structure of water is complex.

Scientists have 'discovered' 36 different types of water, with different combinations of hydrogen and oxygen, and variations of 'heavy' and 'light' water. But these remain in a purely academic realm for most of us. It is convenient but overly simple to describe water simply as $H_2O$; in fact, very little water is $H_2O$. This is simply the 'base' water that picks up minerals on its journey; only distilled water – the pure water of scientists – is actually $H_2O$.

Liquid water is dynamic and chaotic, always on the move in curves and spirals. Its molecules press together in constantly shifting confusion, readily disassembling and rearranging themselves around other molecules. Despite water's ever changing fluidity, its structure is immensely strong.

Water molecules are joined together by hydrogen bonds that are much more adaptable than other chemical bonds. They are strong enough to bind, but weak enough to break easily. These bonds seem to hold the clue to water's behaviour, as they assemble and disassemble millions of times a second, regrouping each time in a very precise arrangement.

Molecules of other substances tend to take on a regular shape, building up with neat and even spacing between atoms. In a water molecule the two atoms of hydrogen always aim to bond with the oxygen atom in a very precise and idiosyncratic alignment – they seek a resting alignment at an angle of precisely 104.5° from each other. The diagram of a single water molecule looks like a Mickey Mouse head, a large circle (the oxygen) with two smaller circles (the hydrogens) in specifically aligned positions. Every water molecule wants to form a three-dimensional structure with its neighbours. Liquid water is made up of billions of tiny crystal-like structures, the more crystalline the structure, the healthier the water.

All early systems of medicine recognized the vital importance of water, not just as treatment for specific symptoms, but also as the source of our life force, the source of health. Greeks recommended bathing in water as a general treatment for the health of the constitution, the Chinese recognized water as the source of *chi*, or life energy, Vedic systems saw water as the source of *prana*, the equivalent of *chi*.

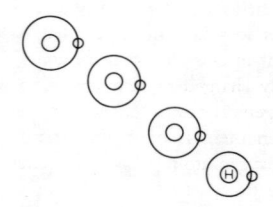

## HYDROGEN

Hydrogen is the smallest atom of all. A hydrogen atom consists of one (positive) proton at its core, and one (negative) electron revolving around it in a three-dimensional shell; it is very light and fluid and always changing.

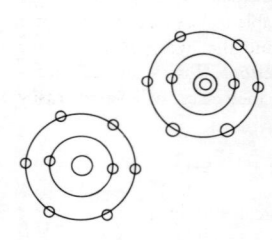

## OXYGEN

An oxygen atom is heavier than a hydrogen atom, with eight protons in its nucleus and eight electrons revolving around it in two outer shells; two filling the inner shell and six in the outer shell. Because its outer shell is incomplete and would prefer to contain eight electrons, an oxygen atom is keen to link up with other atoms with one or two electrons to spare.

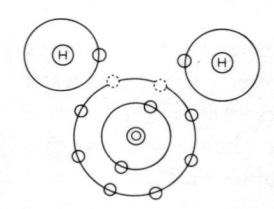

Some people find it easier to understand the way hydrogen and oxygen bond by visualizing the characteristics of each element. Hydrogen can be viewed as a tiny light ethereal body that is very sociable but has a tendency to pull upward, to aim for the skies. Oxygen is equally sociable but tends to head downward toward the Earth. At their meeting point they form the universal mediating element, water, with two light hydrogen atoms bonding to the heavier oxygen. Water is, if you like, the meeting point between heaven and Earth – it forms a cosmic connection.

## H₂O

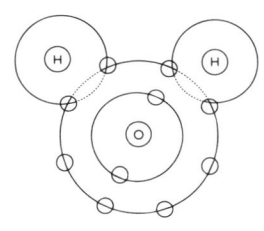

Hydrogen is so small and quick that it can get right in close to other elements. Its positively charged nucleus comes so close to oxygen's electrons that the attraction between them is unusually strong. The larger oxygen atom steals hydrogen's electrons, so this part of the molecule is slightly electronegative and the hydrogen part is slightly electropositive. Although a water molecule is stable, the contrasting charges mean that the oxygen end of the molecule is able to lose electrons to other molecules, and the hydrogen end wants to gain electrons. This structure means that water molecules will bond easily with others.

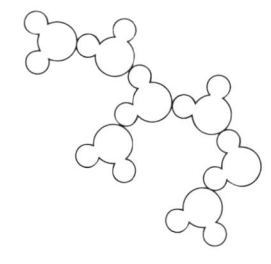

The chemical bonds between water molecules are crucial to its behaviour. If hydrogen bonds were just a bit stronger, water would be solid at 100°C (212°F), then all life on Earth would cease to exist.

# Density of Water

Liquid water is unique because it expands when it freezes. This ability means it can help to preserve life. If cold water was denser than ice, ponds and lakes would freeze (as other liquids) from the bottom up, but when water freezes it floats instead of sinking. A sheet of ice protects the lower waters of lakes, allowing fish to thrive and life to continue; it protects the flowing waters of streams and rivulets. Winter water can lie protected by its own solid phase.

The laws of physics dictate that the solid phase of a substance will be denser than its liquid state, and something solid will sink in liquid. But water is different. Like all other liquids, it shrinks as it cools. But as it reaches 4°C (39°F) something unexpected happens. At 4°C (39°F) water reaches its maximum density – and contains its maximum energy. Below 4°C (39°F) it begins to distend, its density diminishes instead of increasing.

As temperatures drop, expansion continues, and as water freezes into solid ice it increases its original liquid volume by 9 per cent but it becomes lighter than liquid water. A liquid normally becomes denser as it cools, its molecules slowing up and crowding together as the liquid shrinks. But ice is an open, porous structure with fewer molecules in the same space. Instead of closing ranks as they cool, water molecules open outward. They link in delicate structures and latticework, light and buoyant on the water below.

At its densest, the structure of water is tightest; the temperature of 4°C (39°F) allows hydrogen bonds to keep oxygen atoms spaced in the densest possible structure, with water molecules clustered into three-dimensional structures that resemble prisms or hexagonal cages (hexamers) – like two tetrahedrons together; these structures are central to water's behaviour. As water temperature drops, some of the hydrogen bonds, having already reached the temperature for optimum density, are forced to regroup in order for water to expand. While trying to maintain their angle of 104.5 degrees to the oxygen atoms, the pressure causes the bonds to break and reform water molecules with a more open and lighter structure.

The mystery of the Bermuda Triangle has puzzled observers for centuries. Ships and planes have disappeared without trace, leading to all sorts of bizarre explanations. It seems likely that the secret lies in the water. At its densest, water forms strong cage-like structures that, at specific temperatures and pressures, can trap other molecules inside. Dense, and therefore highly structured, water sinks to the bottom of the ocean where it traps methane from the ocean bed. An underwater landslide can cause the methane to be released in an uncontrolled blowout that erupts on the surface without warning. Here, the water would instantly become much less dense and any ship would plunge into the depths to be covered up as sediment disturbed by the blowout settled back on the seabed. A plume of methane gas would continue to rise since methane is lighter than air, condemning an aircraft to a similar fate. Either its engines would fail through lack of oxygen, or the methane would explode when it met hot engine exhausts. Debris would sink rapidly in the low-density water beneath it, leaving no trace.

# From Solid to Liquid

It takes a great amount of energy to turn solid water (ice) into liquid, since this involves changing a liquid from a lighter to a denser form, disrupting the hydrogen bonds in ice. Whereas the bonds in liquid water are constantly breaking and re-forming, researchers have discovered that a high proportion of the bonds in ice, water's crystalline form, are stronger.

If every single hydrogen bond had to be broken, the required energy would be even greater; in fact, liquid water in the form of melted ice retains some of the crystalline characteristics of ice, and one third of its hydrogen bonds remain intact.

The healthiest water of all seems to remember its life as ice, when its structure was light and open, and therefore most open to other influences – elements or vibrations. The purest, healthiest water of all, such as Hunza water, is ice-melt.

If water was able to change its temperature with comparative ease, organic life on Earth as we know it could no longer exist. The lowest specific heat of water, or the temperature at which the greatest amount of heat or cold is needed to change water's temperature, is 37°C (99°F). This is just below normal healthy human blood temperature. Since our blood is up to 90 per cent water, this ability of water to resist thermal change allows us to survive relatively large fluctuations in temperature. If the blood in our bodies had a much lower specific heat we would cool or heat up much more rapidly, to the point where we would either start to freeze or to decompose if exposed to extreme temperatures! The boiling point of water is higher than almost all other liquids; this is due to the massive energy required to disrupt the hydrogen bonds from their stable arrangement, transforming some of them to vapour.

Water's thermal inertia also explains why the climate on our planet is relatively mild and comparatively uniform. Towns near coasts typically enjoy cool summer breezes and mild winters. Oceans store and distribute vast amounts of heat, shifting it from the equator to the shores of continents. An ocean current 160 kilometres (100 miles) wide can transport as much heat in a single hour as could be gained from burning 200 million tonnes of coal.

# Surface Tension

If there is nothing else to grab on to, water sticks to itself. If it is not contained it always wants to return to a sphere. Even rivers and streams, pulled to the earth through gravity, will naturally form oxbows which will then become circular and detached from the main watercourse if the flow is slow. A drop of water is the nearest thing to a sphere that water can become on Earth. But in space, where there is no gravity, things are very different.

One incredible film from NASA, shows astronauts aboard an orbiting space shuttle drinking orange juice from a plastic bottle. Then they gently squirted the juice out of the bottle, stopped squeezing, and the juice came back into the bottle, the liquid returning to stick to itself. When they squeezed hard the liquid flew right out of the bottle into the cabin, where it began to coagulate, drop by drop, into a perfect sphere – because there was no gravity. Then the astronauts began to blow the sphere of juice backward and forward like a game of catch. The film ended with them putting drinking straws into the orange sphere and making it disappear.

Molecules at the edges of any quantity of water are strongly attracted to each other, and cling together in a flexible coat. This elasticity allows a film of water to bulge and stretch in bubbles and drops. It allows blood to clot. The surface tension draws biochemical compounds to concentrate near liquid surfaces, so speeding up biological reactions.

The surface tension, combined with the stable structure of water, gives liquid water an adhesive capacity that allows it to grasp at everything without breaking its structure. Water creeps through cells in tissue. It can even climb upward via capillary action, as in the thin tubes in the stem of a plant. As the fast-changing hydrogen bonds break and reform millions of times a second, most of the molecules attached to the inside walls of the tubes cling to each other and the tube's walls, pulling themselves up. This capillary action is at the heart of all biological life. Without it, circulation would stop.

> Try floating a needle on the surface of a saucer of water. Then repeat the same process several times using water from different sources – you could try seawater, freshwater from a river or spring, bottled water or tap water. You may see some marked differences. You can also try floating objects in hot and in cold water, and see how temperature affects density and surface tension.

## Universal Solvent

Suppose you have been gardening, and your hands and clothes are filthy, covered in soil and general grime. What do you do? You wash them, of course, probably using a little soap along with water. But imagine if you had no access to water, how much energy would it take to remove the grime then? Even if you had soap, it wouldn't help much. Water changes everything. Sometimes called the universal solvent, it dissolves almost everything it meets to some degree – its molecules interfere with every other molecule they meet, pushing atoms apart, surrounding them, and thereby changing them.

The structure of water provides one clue to its amazing ability to dissolve compounds. In order for substances to be dissolved they have to be encapsulated, and this is only possible if the molecules of the liquid form a three-dimensional structure. If water molecules arranged themselves in flat or linear structures, they would not have the adaptability to dissolve anything.

The other factor allowing water to dissolve most substances is its electrical polarity – the electropositive hydrogen atom seeks to attach itself to any negative ions it encounters and the negative oxygen atom then looks to claim any positive ions to keep in balance. Every living creature on Earth makes use of the powerful solvent properties of water; many of our biological reactions are switched on or off by bio-electrical changes, or changes in the concentrations of dissolved ions such as sodium and potassium in the watery fluids in our bodies. Throughout our planet, healthy water is in a constant state of motion and transformation, creating and re-creating as it collects and transports oxygen, nitrogen and carbon dioxide from the air, and calcium, potassium, manganese, sodium and other minerals from the stone. Its extraordinary power to dissolve whatever it touches, and its continuous dynamism, means that it is continually collecting new information and depositing it elsewhere. Water is the place where other elements meet and become transformed. Water structures our world.

If a drop of water touches a crystal of salt the salt will dissolve almost immediately. This is something we all take for granted. But if we tried to melt table salt without water we would have to heat a furnace to a temperature of around 800°C (1472°F)! Yet salt melts effortlessly when it comes into contact with water. This is because sodium and chloride ions are constantly looking to attach themselves to molecules of other substances. When in a solid state the two are very stable and strong as the negatively charged chlorine complements the positively charged sodium. But as soon as they are surrounded by water they are off, the chlorine attracted to the electropositive hydrogen, and the sodium to the negatively charged oxygen. So a solid salt crystal dissolves in a flurry of charged atoms. For this reason salty water will conduct an electric current, although neither water nor salt will do so without the other.

# A Vibrating World of Information

We need to know about the structure and the physical properties of water to help us to put together the wider picture. Each of water's roles and functions is as important as the others. All aspects of water are related and support each other; none can be fully understood in isolation. Whatever you add to water is taken in by it; whatever you do to water, it will then do to you – if we pollute our water it will then pollute our bodies (as well as our environment) and cause ill health. Pollution may be electromagnetic as well as chemical.

Our perception of the world is gradually changing, altering our understanding of everything within it. Some disciplines are moving toward – or returning to – a multidimensional view of the world, where everything is interrelated, with the individual connected in some way to the wider universe. This perspective suits the study of water, the ultimate holistic substance that affects us all on the physical and the emotional level.

The holistic psychologist Carl Jung equated water with our emotions, as the medium through which subtle information is transferred.

A new understanding sees all life's processes as dependent upon energy. Even apparently dense physical objects are coming to be seen not as stable and solid but considered to be made up of vibrating energy. Every object and every substance, whether natural or manufactured, has its own vibrational pattern. Every living thing consists of energy vibrating at a specific frequency.

The overriding importance of energy and energy balance has been understood for millennia in China where energy, or *chi*, is the force underlying all life's processes, the basis of their understanding of life, nature and the body. *Chi* arises from the interaction of yin and yang, the dynamic out of which life and all phenomena arise and continue to move and change.

The simple example of water boiling in a pot sums up the way matter and its changing state are characterized by its energy, *chi*. Liquid water is yin; is ultimately vibrating energy; as it boils, using fire (yang), it transforms to steam (yang) which then may condense into droplets (yin). The most yin form of water is ice. Water is constantly transforming into yin and yang states, creating the energy transformations upon which all life depends.

For centuries science has been bounded by a materialist and mechanistic view of the world – a Newtonian perspective. This sees the physical world as separate from the cosmic, and all matter as fixed and solid, made up of a collection of empirically identifiable chemicals in particular combinations. Newer science is working to a different model, based on principles suggested by Einstein. This view understands the world to be a series of physical forces in dynamic interplay with complex energetic forces. Nothing is solid but all matter is ultimately vibrating energy.

## Resonance

Marching armies always break step when crossing bridges. This is because the rhythm of their marching feet sets up strong vibrations at a specific frequency that might cause the fabric of the bridge to fall apart. This is resonance.

We should recognize the principle of resonance in order to understand how living things communicate via vibrations. Resonance is one of nature's characteristics. It explains the way that different objects in nature tune in to one another, and the way in which energetic information is transferred from object to object through space. Water appears to be one of the most powerful channels for conducting resonance.

If we look at all physical matter as vibrating energy, we can appreciate that different objects vibrate at different frequencies, with some resonating on the same frequency. If you strike the note C on a tuning fork, for example, every C on an adjacent piano will vibrate – all the notes on the piano that recognize that note will resonate with the tuning fork. They may be higher or lower, but they all vibrate. A popular experiment in 'hands-on' science museums encourages children to make different patterns in grains of sand on a metal plate by making different noises, showing how patterns vary according to noises at different frequencies.

Physicists have experimented with sophisticated machines called particle accelerators that shoot intense electromagnetic waves at particular atoms. These atoms simply come apart at certain frequencies as the electromagnetic energy disrupts them. Something similar occurs naturally when an opera singer destroys a glass through the power of their voice, by transferring energy through space to a glass. If the transferred energy is on the same natural vibrational frequency as the glass, energy is transferred to a point where the molecular structure of the glass is under such stress that it disrupts and the glass shatters.

Our bodies are a collection of organs and cells vibrating at or resonating to different frequencies. Or we can see a body as a complex system of harmonic frequencies, and any change in the harmony is the cause of pain, discomfort and disease. Water seems to be the clue to resonance, the channel through which all living things are able to communicate with others. Water stores and transfers frequency information, it mediates vibrational energies from the environment.

# The Memory of Water

The vibrations in water depend on its three-dimensional structures which instruct and inform all life's processes. These microstructures form and dissolve millions of times a second. They can be seen as vibrating centres of energy, constantly receiving and transmitting energy from or to all they come in contact with.

Despite its fluidity and its ability to metamorphose into different states, water molecules tend toward a stable structure. This gives them the ability to store information obtained from other molecules. When you drink clear, lively water from a mountain spring you taste the minerals that the water has brought to life from deep beneath the ground. But you also drink water's vitality, receiving energy through participating in its journey.

As water travels it carries physical information, but it also picks up more subtle information. It responds to every change in its surroundings by expanding, contracting, or making rhythmical waves, changing billions of times a second. And as it changes it not only picks up physical information, but also vibrations. Water transmits vibrational as well as physical information; it affects and is affected by everything it meets, and it brings this information with it to every fresh encounter. We may always have intuitively known about this ability, but we have only recently been able to explain it.

> A snowflake is made up of a quintillion molecules of water organized in a specific and stable arrangement. You will never ever find two identical snowflakes, even if you spend all your life trying. But if you melt a snowflake, and refreeze it under the same conditions, it will refreeze into exactly the same pattern, not a similar arrangement, but precisely the same one! It remembers its previous arrangement, and goes right back to it.

To a non-scientist, the idea that water has a memory is not particularly astonishing. Every living thing is affected by its environment, physically and emotionally. Human beings retain impressions of all that happens to us on our journey from life to death. Some of these impressions are heightened, if they were particularly positive or negative, some are hidden deep in our subconscious; every human being is affected by every one of their experiences and by the total of them all. So if we recognize water as a living entity it should be no surprise that it has a memory.

If you pour water from a jug into the cupped hands of two different people, you instantly obtain three different kinds of water. The water in the jug contains one set of information, but that information is changed or added to through contact with the hands of its recipients. As water changes everything it touches, dissolving and freeing elements from their previous states, so everything changes water. Water is affected by the chemical information contained in a person's hands, for example – what elements or traces of elements may physically be present on the skin – but also by the very pulse of that person, their rhythmic patterns and vibrations, and vibrations that it picks up from the cosmos, from the moon and from planets.

# Scientists and Water Memory

The idea of the memory of water has raised some extremely controversial issues among scientists. A French biochemist, Jacques Benveniste, was at the forefront of the arguments in the 1990s, although other researchers have been working on the subject. He showed that molecules of other substances use water as a medium for communication, and that water acts as a transmitter of stored physical and vibrational energy. Benveniste hit the headlines because he was respected as a mainstream scientist; most other work on the memory of water had been done on the sidelines of conventional science, and was therefore easier to ignore. Unable to ignore him, the establishment instead derided his methodology and his results, and accused him of fraud.

The idea of water transmitting vibrations is not new. The practice of potentizing water by exposing it to subtle energy, or to specific electromagnetic frequencies, has been practised for around 200 years in homeopathy, where remedies charged with different frequencies (at different resonances) produce different biological effects. But homeopathy has never been fully accepted by either conventional science or orthodox medicine.

Benveniste's experiments observed quantifiable effects on living cells using accepted scientific methodology. They showed that water does retain and transfer information, and its molecular structure allows it to act as a transmitter and receptor of both chemical and vibrational information. According to his work, reactions occurred at dilutions of one part in 10,120 of distilled water (1:1 with 119 zeros!). This is equivalent to adding a drop of a substance to 99 drops of distilled water, then a drop of this mixture added to 99 drops of water, and so on 120 times. Benveniste showed that these extreme dilutions can modify biological activity in water – even though no molecules of the original substance are present. His results can be interpreted to indicate the memory of water, how water retains the imprint of the image and the original qualities of the preparation.

If water is passed through a magnetic field, many of its properties change, including its boiling point, viscosity, surface tension, electrical conductivity and magnetic susceptibility. The water does not revert to its previous state. Moreover, it seems to have the ability to affect any other water it comes into contact with. It remembers it has passed through a magnetic field, and passes that information on to other water.

# Homeopathy

Homeopathy is a safe and gentle form of healing that acts on the principle of 'like cures like', the idea that anything that has the power to harm the body also has the power to cure it. This ancient principle was formulated into a specific discipline by Samuel Hahnemann nearly 200 years ago in Germany.

Homeopathic remedies are prepared by shaking, or succussing distilled water (i.e. pure water containing no traces of any physical substance) in order to transfer a mineral or organic substance to the water. These preparations are then diluted many times, so that no trace of the physical substance remains in the water. What does remain in the microstructure of the water is the vibrational imprint of the substance.

Homeopathic preparations are effective because water is able to extract and store certain types of subtle energies that have measurable effects on living systems. The remedies work directly on the energy frequencies within the body.

Hahnemann discovered the importance of succussion by one of those accidents for which science is known. In his early experiments with remedies at high dilutions he found that the most effective preparations were not those that he gave to his patients in his consulting rooms, but those that he personally delivered to patients. So he deduced that this effectiveness must be due to the rhythmic shaking the preparations experienced on the journey by horse! We now recognize that movement is an integral part of water's energy and memory (see Chapter 3).

Since there is no physical substance left in the preparation, a homeopathic remedy does not act directly on the physical body, but on a different, more energetic, level. Hahnemann referred to this level as a person's vital force, the energy that tries to keep the body as healthy as possible. Usually, the body's vital force copes with most problems, but there are times when it needs help – for example, clearing out toxins and rebalancing itself during a common cold; or fighting infection and burning up impurities during a fever; or keeping problems away from the vital internal organs during skin diseases.

If homeopathic remedies are stored near equipment that gives off electromagnetic vibrations – electrical motors, microwave ovens or computers, for example – the information they contain can be subtly altered as the electromagnetic fields can destroy or change their vibrational patterns.

# Influences on Vibrations and Memory

We live surrounded by different energy fields, some of which have obvious or measurable effects, others less obvious or not apparent at all. Before delving further into the significance of water memory, it is worth looking at these fields. A field refers to an area of space that has particular properties. Magnetic and electromagnetic fields subtly change the frequencies at which things vibrate.

We all need the effects of the naturally occurring energy fields that surround us. Problems occur when the fields are altered, usually through technology, and the natural balance of the environment is lost.

## Electric Fields

The Earth is surrounded by an electric field, which extends from the ground to a height of around 160 kilometres (100 miles) in the upper atmosphere. An electrical tension exists in this field that varies over a range of frequencies, but stabilizes at approximately 7.83Hz – similar to the frequency at which some human brain waves resonate.

The Earth's electric field is relatively stable and has little or no effect on us normally; we rarely notice it. We start to be aware of it during the build up of a thunderstorm when the electrical tension between atmosphere and Earth is changed, thereby changing the strength of the electric field. Locally, the field can be affected by the electromagnetic frequencies emitted by some equipment, such as around power stations and cables.

The most dramatic evidence of the Earth's electric field appears in a lightning strike, shooting an electric spark from the atmosphere to the ground. You can screen any object against an electric field by placing it in a metal cage (known as a Faraday cage). Hence, the safest place to be in a lightning strike is inside a car with the windows firmly shut.

Every raindrop that falls during a thunderstorm collects a charge from the atmosphere and deposits it on to the Earth. Positive ions predominate, and these make most people – and living creatures – ill at ease or agitated. After the storm has passed, beneficial negative ions take over, with a more relaxing influence.

# Magnetic Fields

The Earth is like a huge magnet that generates a magnetic field that extends far out into space. Like the Earth's electric field, we are rarely aware of it yet it constantly passes through us. Some living organisms are attuned to this magnetic field: birds, for example, seem to be guided by it in their patterns of migration and navigation, and can be confused by even small changes in its strength.

The magnetic field is affected by dense currents of charged particles emanating from the sun on the solar wind. You can witness this effect in the startlingly beautiful displays of the aurora borealis around the magnetic north pole and the aurora australis around the magnetic south pole.

To a much lesser extent the Earth's magnetic field is influenced by planetary movements, too. The magnetic field can also be affected by strong storms – a massive surge of positive or negative ions, for example, would create a powerful local magnetic field. Magnetic rhythms are part of life and they penetrate the organs and tissues of all living organisms.

The Earth's crust carries a positive charge beneath it and a negative charge on its surface. A positive charge in the upper atmosphere emanates from the sun's energy. We need the negative field on the Earth's surface for balance and health, but we also need the dynamic of the positive fields above and below. The south pole of a magnet is positive, and increases and expands the energy of whatever substance it is applied to. The north pole is negative, and reduces and dissolves energies. A negative magnetic field is appropriate to dissolve blockages in the body, pulling energy down and grounding it. Positive magnetic energy will magnify a person's energies.

# Electromagnetic Fields

Electricity and magnetism are inextricably linked. In a domestic power supply, the electricity is conducted along straight courses around specific circuits, so it vibrates at a slow and steady level. But the electricity that occurs in the environment (in certain meteorological conditions, around power cables, etc.) vibrates at a wide range of frequencies. When electricity vibrates it turns into magnetism – for example, storm conditions release electrical charges that create magnetic fields. Similarly, when magnetism vibrates it transforms into electricity – scientists such as Michael Faraday and Nikola Tesla used this knowledge as the basis for generating electricity, setting magnetic coils vibrating to create electric charges.

We are not sure exactly how this transformation works, we just know the connection exists. Dynamic and vibrating electric or magnetic fields become electromagnetic fields constantly vibrating at high frequencies and spreading out into space. Electromagnetic fields, or EMFs, are often cited as 'new phenomena'. But they have always existed, on various frequencies, coming from the interaction of the Earth, the atmosphere, the ionosphere, the sun and the planets. The sun's radiation is electromagnetic energy. A flash of lightning creates electromagnetic fields.

For the purposes of simplification, low frequency fields (e.g. those given off by mains-powered household electrical equipment) can usually be seen as electrical or magnetic. High frequency fields are usually electromagnetic. It is possible to transmit information over distances even at low frequencies, and most low frequency emissions normally encountered are harmless. However, extremely low frequency waves, known as ELFs, can be dangerous and destructive when transmitted in concentrated volumes. They can damage the autonomic nervous system, almost literally crossing the wires to the brain, altering the information that a body needs to maintain its life force.

We all need a certain amount of electromagnetism in our everyday lives – we all need moderate radiation from the sun to maintain our health. But the numbers of EMFs we are exposing ourselves to is growing with increased technology, such as TVs, microwaves, mobile phones and computers.

EMFs can distort or destroy the structure of water and hence the information that water carries. So we need to be careful not to expose ourselves unduly to EMFs, or the delicate balance of water in our bodies will be subtly changed. Moreover, the effects of exposure may be cumulative – water remembers what it has been exposed to, and relays that information to other water.

You can look at the human body as a mass of electrically charged molecules that move along bio-electrical circuits – electrical currents that cross through arteries, veins and capillary walls, drawing white blood cells and metabolic compounds into and out of surrounding tissue. EMFs affect this electrical exchange and distort the subtle energy that surrounds the body, known as the bio-electromagnetic field. Electromagnetism can alter information so that it destroys, rather than enhances, health. Controlled doses of very high frequency electro-magnetic waves do have application in medicine, as in X-rays, but the dangers of overexposure to X-rays are well known. We can't avoid exposure to EMFs in the environment, but we can minimize the risks.

Just as living in the vicinity of power cables can cause health problems, so exposure to electromagnetic fields

(EMFs) can alter the structure of water. Recent research indicates how underground streams can be powerfully influenced by EMFs, destroying the structure of the water and imprinting new vibrational information on to it. People living in the vicinity of such streams affected by EMFs can suffer ill health. Such influences can be healed by natural means such as crystals, plants (e.g. cacti) or specially designed devices.

Dowsing is a bridge between the logical and the intuitive mind. Dowsing, or water divining, is a means of locating concealed sources of water. Animals can sense it, Māoris just use their feet, others 'hear' it. The difficulty lies in finding the kind of water you need. In 1518 Martin Luther stated that the use of the divining rod broke the first commandment. Even today some people see dowsing as the devil's work. Dowsers may be sensitive to electromagnetic frequencies given off by water, but research by Britain's senior astronomer, the Astronomer Royal, suggests the signal is involved with cosmic radiation. Dowsing is one obvious way in which water can transmit its signals through space. A dowser's tool merely amplifies the message that is in their body as it resonates with hidden water.

# Water, Vibrations and Health

The world is made up of vibrations. Water picks up vibrations in a way similar to an audio recording of a voice; water becomes active when vibrations are imprinted onto it.

Benveniste's experiments showed how biological information contained in water could be transmitted from one phial of water to another using a low frequency electronic amplifier. Particles in the water can therefore act as radio transmitters, picking up vibrations, or electromagnetic information, from other particles. The information was destroyed when the water sample was subjected to a strong magnetic field, or heated above 70°C (158°F). At this point the structure of water changes, bonds break, and water can no longer transmit information.

> If you put a magnet in front of your mouth it doesn't stop you speaking, but a magnet will distort a tape of your recorded voice. Recorded information consists of electromagnetic vibrations, which are affected by magnetism. Information that is homeopathically imprinted on to water is also distorted or eradicated if it is subjected to strong magnetic fields. So information in water also consists of electromagnetic vibrations.

Simpler experiments were carried out by another scientist, Vitold Bakhir, in Russia. He placed some water in a petri dish, put a glass plate on top of it and placed another dish on top of the plate with a closed phial of 1% sodium chloride in it. The next day he returned, froze the water in the petri dish, and analyzed the resulting crystals. They contained sodium chloride, which had transferred through the glass from one sample to another.

> Put a bottle of water (plastic or glass) in your refrigerator next to a ripe melon. After 24 hours pour a glass of water from this bottle and ask a friend if it tastes of anything in particular. The water will probably taste of melon.

Every chemical element has its own vibratory pattern, whether dissolved in water or not. This is a kind of code, to be picked up by an appropriate receiver. If water is healthy, it can pick up and transmit these codes. Water acts like a radio receiver and transmitter: without the waves for communication a radio can't do anything, but when it picks up appropriate signals it can communicate. So when a molecule of any substance meets water, its potential for communication increases a thousandfold. A radio transmitter can't pick up signals if they are blocked by physical obstacles or other radio waves; water can't pick up the correct codes if blocked by pollutants that scramble the codes. If molecules only communicated when they bumped into each other, biological communication would be clumsy, or would break down completely.

Imagine an aeroplane arriving at JFK Airport in New York, and suppose that it could only communicate with controllers at JFK when it was near enough to see the control tower. Interaction would be difficult and ineffective. In fact, the plane is able to pick up information on a specific frequency long before arrival. Water similarly allows molecules to communicate without touching each other, via their specific vibrations. Thus water has the ability to allow molecules to interact within a cell and to conduct molecular messages so that they reach different parts of the body via the watery fluids in intercellular space.

In Hunza water, minerals are present suspended in minute clusters called colloids. Colloidal minerals have perfect electrical and magnetic balance, so Hunza water has magnetic as well as electrical energy. Ancient and fossil waters contain traces of colloidal silver. Silver is an antibacterial element, currently used in some water filtration systems.

# Water Structure and Vibrations

But how do the cells pick up information from water? Are our bodies just governed by chemical reactions? A living human being, or any living thing, is more than a combination of chemicals. But a dead body is simply a physical object degrading into a disordered collection of chemicals. The difference is the life force, the energy, *chi* or *prana*. This comes from vibrations.

Healthy water has a strong three-dimensional crystalline microstructure that allows it to collect and transmit information. It is the structure of the water that allows it to communicate. (This is equally true for cellular water, extracellular water, drinking water or water in the environment.)

Cells communicate via vibrations – a cell is a bio-electronic factory, oscillator and miniature radio receiver. Healthy cells vibrate at specific frequencies. Life depends on these vibrations that allow the cell to pick up the appropriate resonances from the water that enters and travels round our bodies, and to pick up the energetic messages from healthy water. Blockages to the free flow of energy through the body lead to illness.

Water's health and energy change with movement, temperature (when water picks up different elements), electronic exchange, or when water receives electric, magnetic or electromagnetic vibrations. Water acts on every level of the human being. You can ensure general health simply by using good quality water with a healthy structure and good energy; or you can look to therapies using knowledge about the vibrational energies of water and the energies of your body.

# Healthy Water for a Healthy Body

We are largely made of water and it is the channel for all physical and chemical changes in the body. A newborn baby is about 97 per cent water, a healthy adult 75 per cent, and as we get old we typically dehydrate to become about 65 per cent water. A brain is around 75 per cent water, bones 22 per cent water, even tooth enamel is about 2 per cent water.

Our bodies are a mass of finely tuned chemical and biological reactions, relying on the remarkable properties of water for life. We depend on water as a catalyst, a transport system, to maintain our body temperature, and as a supplier of nutrients or electrical impulses.

The body is a collection of trillions of cells, separated and filled by watery fluid, and this fluid balance is the key to health. A person can exist for several weeks without eating any solid food, but without water they dehydrate and die within days. Even a 2 per cent loss of the water that surrounds our cells (extracellular water) can mean a 20 per cent decrease in energy levels. We need to drink about eight glasses of water a day in order to maintain health. Alcohol, tea, coffee and fizzy drinks are diuretics. They do not add to your fluid intake but deplete it. If you drink too much of any diuretics you will need to drink more water, rather than less.

If we don't drink enough water we dehydrate. If we dehydrate we are likely to suffer all kinds of degenerative diseases. We should all drink 2 litres (3.5 pints) of pure filtered water a day. Few of us have access to entirely healthy, unpolluted water. But nor are we likely to suffer serious health problems from drinking tap water. If you are doubtful about the quality of your drinking water and have no access to filtration systems, simply add a teaspoon of Vitamin C powder per litre of water. This will help ensure the water maintains an appropriate level of acidity and make it easier for your body to assimilate.

There is nothing remotely controversial about this, our health depends on the electronic exchange, or exchange of energy, in our bodies. This exchange is dependent on water – and on the health of the water we consume: the water in and around our cells must be the right polarity to catch the appropriate ions of sodium, potassium, magnesium and others which nourish the interior of the cell so that we can function. The polarity of water depends on the positive ion exchange of minerals in the water. The potassium and sodium balance in the body is finely tuned, the two elements establishing a crucial dynamic tension. Depletion of either would reduce the ability of a cell to respond, affecting amongst other things the vital acid/alkali balance.

A healthy cell takes in nutrients from the water outside the cell, and the inner and outer cellular water is balanced through a process of osmosis via the membrane of the cell wall. The cellular water that surrounds the cell should be of weaker concentration than that within the cell, so that the outer water can remove toxins from the inner cellular water. If drinking water is too full of electrolytic conductors – excess of certain minerals as well as toxins such as heavy metals which can't be assimilated by the body – this will affect the polarity of our extracellular water and can even end up reversing the process of osmosis. In extreme cases the polarity of cells can become distorted or reversed through this process, and so diseases such as cancers may occur.

The health of our body cells is maintained by a constant process of renewal. Anything that interferes with this process leads to cellular deterioration, that is, ill health and ageing. It is important to drink clean and healthy water as the more complex the pollutants in the water, the more difficult become the cleansing, cell renewal and transport of nutrients in our bodies. There is a direct correlation between the degree of water cleanliness, efficient human cell regeneration and ageing. One ancient Indian technique, Kaya Kalpa, stops, or even reverses, the ageing process through fasting and drinking healthy water to facilitate the physical cleansing that supports the subtle meditation practices.

---

## HERE ARE SOME GUIDELINES FOR HELPING CHILDREN GET INTO THE WATER HABIT:

- Don't let your children get into the habit of drinking processed fruit or fizzy drinks.
- Always have a jug of filtered water in the refrigerator, or install a filter for a special drinking water tap by your kitchen sink.
- Discourage children from adding ice to water – drinking water that is too cold can shock the system.
- Rather than buying concentrated fruit squash or juice concentrate, encourage children to add a slice of fresh fruit to a glass of water.
- Help them to develop healthy drinking habits.

# Subtle Energy Exchange

Water does not only pick up and transmit molecular information, but also extremely subtle energies. In the 1960s, a Canadian researcher, Bernard Grad, performed many experiments on healer-treated water. He proved that water held by a psychic healer had a positive effect on the growth of seeds, and also proved that the water that had been in contact with a psychologically depressed patient had the reverse effect. So water was able to transmit both positive and negative subtle energies. When Grad scientifically analyzed the healer-treated water, he found that the hydrogen bond angle of the water molecule had undergone a subtle but detectable shift.

Water can affect human beings without even touching them. At a conference in the UK a Swedish researcher told of remarkable observations about how the water from some Swedish mountain springs can have a healing effect on individuals without their drinking it. Measurable short-term healing effects were obtained when patients were merely in the presence of the water! This miraculous water, when analyzed, will probably be found to have an ice-like structure, and contain minerals, or a memory of minerals, from deep in the mountain.

An increasing number of healing practices are evolving based on ancient Chinese and Indian understanding of the need to balance the body's energies. Vibrational healers look to rebalance the energy field (the body's bio-electromagnetic field) that regulates cellular structure and function. The aim is to restore order and harmony on the level of an individual's life force. In esoteric science, medicine or philosophy, subtle energies that mediate the life force surround an individual on four levels: the spiritual, the mental, the astral and the etheric.

These levels are known to vibrate at differing frequencies.

The ancient Greeks viewed ether as the fifth element, the unseen element surrounding and permeating all life, determining the interaction of the other four elements (water, air, fire and earth). Nowadays, we understand the etheric body to be the energy template of our physical body. Physical illness may begin at the etheric level, before cellular physical changes have started. Some sensitive people perceive the etheric body as a halo of colour, or aura, around the physical body; various layers of the aura have been described corresponding to the physical, emotional, mental and spiritual aspects of the individual. The etheric or energy body influences people's emotions, minds and spirits as well as the physical body.

In Chinese medicine the subtle energy, *chi*, flows from the environment to the nerves, blood vessels and deeper organs of the body along special channels called meridians. Indian medicine recognizes seven specialized energy centres, or chakras – each is associated with a major nerve and glandular centre in the physical body. Connected via energetic threads known as *nadis*, the chakras act as transformers to translate subtle energies and relay them as hormonal, nervous and cellular messages in the body. The major chakras – brow, crown and throat – are also subtle organs of perception associated with the psychic abilities of higher intuition, clairvoyance and clairaudience.

> Try a simple experiment suggested by Joseph Bender, a water researcher in Texas. Take some distilled water and pour equal amounts into five clean glasses. Then hold one of the glasses in your hands and project loving and healing thoughts into the water, before asking a sample group to taste the water. In Bender's own experiments, repeated

many times, 6 out of 10 adults correctly identified the water that had been subtly changed through positive thought – and 9 out of 10 children said which was best. Water can exchange energies that science does not yet comprehend nor have the equipment to measure.

## Flower Essences

It is believed that the flowers of any plant contain the highest concentration of life force or energy as they are the crowning experience of the plant's growth. Bach Original Flower Remedies use the energies of sunlight to transfer the essence of a flower onto water, containing energetic or vibrational information from the sun as well as the energy of the plant. Sunlight acting on a plant transfers an electrical charge from the plant to the water – the morning sunlight makes dew so vitally active that walking barefoot in dew is in itself a traditional Indian method of healing the body.

The sun's action in conferring energy on water is known as solarizing the water. The sun's positive electrical and magnetic energy holds the healthy negative electrical and magnetic energy in the water, balancing the fire and water energy, the yang and yin. Solarized water is balanced electrically and magnetically.

Edward Bach was an English doctor who became a homeopath in the 1920s. He developed Hahnemann's work on individuals' characteristics and predisposition to certain illnesses (and therefore certain remedies), and concluded that emotional and personality factors contribute toward general predispositions to illnesses. He believed that the link between illness and personality was a result of disharmony between the physical personality and their Higher Self, or soul, creating dysfunctional energetic patterns.

Bach developed an incredible sensitivity to subtle energies and was able to discover the effects of the various flowers through observation of how they affected him. He prepared flower remedies by placing the flowers of a particular species on the surface of a bowl of spring water for several hours in sunlight. Then the flowers were removed and the water mixed with alcohol to stabilize the essence.

Flower essences are vibrational remedies, where water is imprinted with the 'essence' or 'signature' of a specific flower to treat a variety of emotional disorders and temperaments. Flower essences affect the physical body via the emotions, and seem to be particularly potent in causing changes in the chakras and meridians, while homeopathic remedies deliver frequency- specific vibrational information that seems to resonate more with the physical/ molecular structure of the cellular body. So Bach remedies go to the emotional root of problems, and homeopathic remedies to the physical root. When any vibrational essences are taken internally their energies are amplified and assimilated as they resonate with the structured water in our bodies.

The 38 Bach Original Flower Remedies are widely used by naturopaths and other holistic practitioners. Mimulus is given for fear; White chestnut seems to help against obsession; Scleranthus against indecision; Star of Bethlehem can be taken for shock. Rescue Remedy, a combination of various flower essences, is widely used to combat any stressful situation.

# Directing Healing Energy into Water

The principle underlying all vibrational methods of healing is resonance. The molecules that form the tissues of our bodies are charged electrical entities and many of the diseases affecting our bodies are primarily due to an electrical deviation from the norm in the structure of the molecules. It is considered that each molecule, of which all organs and tissues are composed, has its own individual sound pattern and emits a vibration peculiar to this pattern; in a healthy organ the molecules work together in harmonious relationship with each other. Any disruption of this harmonious relationship constitutes disease. One vibrational field of medicine, cymatic therapy, aims to restore a body to physical harmony so that it resonates on its own appropriate frequency. Diseased cells exhibit different frequencies to healthy ones, but can be returned to balance by bombarding them with sound waves.

Cymatic medicine works on the principle of resonance. Picture an orchestra with all its members are playing Beethoven's 'Ode to Joy' apart from one sole violinist who has decided, come what may, he is going to play a nursery rhyme. The rebel may be able to play this for several rounds, but he will be so overwhelmed by the sound the group is making that he will be unable to continue; he will be, quite literally, drowned out by the power and resonance of the mass and will be forced into conformity – or silence if he does not know the other piece. This is what happens in cymatic medicine: diseased cells cannot cope for long with sound that is directed at them and which resonates on the same frequency as that of the healthy cells. So they eventually begin to resonate on the healthy frequency.

Aquasonics, developed by Peter Guy Manners in Worcestershire, England, uses this principle as the basis for its healing techniques, with water as the medium to transfer sound vibrations to a patient. It treats affected areas of the body with natural frequencies of sound, working in harmony with the natural processes of the body.

In order to transmit the sound vibrations, Aquasonics introduces specific frequencies into a bath or pool of pure mineralized spring water, free of chemicals (and therefore free of other vibrational information). This water is first structured and energized through an electrochemical process so that it can trap the frequencies in the water's energy. Vibrations are introduced to appropriately structured and energized water; they can remain effective in a pool for up to 12 weeks. We are beginning to have access to a number of different methods of energizing water, and some modern techniques are suitable for home use (see Chapter Three).

The practice of healing through sound is not new, but originates in the Chinese systems of medicine. Healing with sound and water also appears in more recent agrarian societies, such as in a practice called *tonsingen*, which directs specific sounds into water to energize and restructure it.

# Crystals and Colours

Every plant, animal or mineral has an aura, or bio-electromagnetic field. This aura oscillates at a specific frequency, and is affected by the fields with which it comes into contact. This is how organic and inorganic objects (i.e. cells and crystals) communicate with each other and become part of a whole unified energy system.

It is no coincidence that the healthiest water has a strong crystal-like structure, able to collect, transport and transmit energy wherever it flows. Crystals are powerful, they have been used for millennia to draw energies from unseen worlds to heal environments and people. Ancient Egyptians, for example, not only built pyramids to harness earthly and cosmic energies, they also pulverized gems, mixed them in liquid and drank the potion as a medicine. Modern gem elixirs once again recognize this power, crushing specific minerals in water to soothe a variety of ailments. Crystals can heal and energize water, and they can heal individuals.

Crystal healing is sometimes referred to as a 'passive resonance therapy', like cymatic medicine. It also works on the principle that every cell vibrates at its own specific frequency, with crystals able to balance these frequencies, restoring them to their appropriate levels. Just as water gives back to us what we put into it, our relationship with crystals is also crucial – they vibrate at a natural healing frequency that appears to depend on intention. Crystals seem to direct healing energy to all who seek their help.

Since water is the most effective carrier of vibrations, any crystal healing can be doubly powerful if these two elements are combined. To restore flagging energy levels, enjoy a soothing bath with crystals placed at points in the bathroom. Or add some Aura-Soma remedies to your bath (see page 159) – these combine the energies of crystals, plants and colours to balance your body's bio-electromagnetic field.

Colours also have energies. Each colour vibrates at a specific frequency, and different colours relate to different physical and emotional needs. If you are anxious or stressed, lavender, for example, vibrates on a level that helps balance your emotions and your physical needs, drawing calming and soothing energies down into your body to help produce a general feeling of wellbeing and boost the immune system. Some people like to drape appropriately coloured cloths over water bottles to help their water maintain its energy; others bathe in coloured water and absorb the light frequencies over their whole bodies, getting maximum benefit from the healing energies of colour.

# Chapter Three
# Using Water's
# Vital Energies

Energy is integral to every living thing. Human beings' energy levels vary hugely – sometimes we may feel extremely energetic, at other times sluggish or even exhausted, scarcely able to contemplate the simplest action, preferring to curl up and sleep. At death there is no energy left. Water is no different. In some situations it is full of energy, sometimes it is sluggish and exhausted; in some conditions it dies. Whereas all other living things can be seen to decompose after they die, turning into a heap of disordered chemicals as their life force disappears, to an untrained eye dead water still looks like water. So we still expect it to do the same job as healthy energetic water.

If you get the chance, take a walk up a mountain and visit a mountain spring. Look at the water, listen to it, taste it, experience it. Compare this experience with that of turning on a tap and filling a jug with water from a normal domestic supply. Or stroll by a river where water tumbles over boulders and rocks, swirling and eddying on a meandering course toward the sea. Then visit a municipal reservoir where the water is dammed and contained by cement banks and sluice gates.

All is water, but qualitatively – and quantitatively – it is very different. Bubbling spring or brook water is not just different because it contains high levels of certain minerals; it has a vital quality that is lacking in the other waters. The still reservoir water, on the other hand, that feeds the mains supply to our households, is likely to be almost devoid of energy, and therefore unable to communicate.

The way water works in nature is reflected in the way it works in our bodies. A flowing stream is self-regulating, self-cleansing and healthy, with no algal deposits; while a stagnant pond is full of all kinds of bacterial growth. In the same way a healthy body has the correct bacterial balance and can heal itself.

If you add a fountain to a pool of still water the water will become cleaner, livelier, healthier and more energetic; the balance of bacteria and micro-organisms will be restored. This is the same process that occurs when a body receives healthy water – it can function properly as all the appropriate biological responses are activated, while it is much harder for the body to assimilate and use unhealthy water for balanced health.

We don't all have access to highly energetic fresh, healthy, water for all of our daily needs. So make the most of it whenever you come across it. Take a walk by fresh water as often as you can, dip your feet into it, splash your face with it. And if you know of a clean source of spring water fill a glass bottle and take some home with you. You can even bring some energy back to your tap water by stirring it into spirals, or by running some into a jug and leaving it for a while in a cool shady place before using it for drinking or cooking.

## Viktor Schauberger and Living Water

How can we ensure that water is healthy? One of the earliest commentators on the state of a nation's water was Plato. This ancient Greek philosopher warned of the consequences of deforestation for the health of water: as trees are felled shade vanishes, soil erodes, watercourses change as water runs into them from the eroding land, water becomes full of sediment, sluggish, unable to follow its natural course.

At the beginning of the 20th century, a young Austrian forester, Viktor Schauberger, paved the way for much of our modern understanding about the vital energies of water. Schauberger is often referred to as the 'Water Wizard of Austria'; his insights came from observing living water: the way water meanders from side to side, the way water snakes undulate through water and the way cool water can even carry stones on its surface.

His central theme was that current technology used the wrong kind of motion: machines and industrial processes tended to channel agents such as air, water and other liquids into straight lines. These cause destructive pressures since they are motions that nature only uses to decompose and dissolve matter. Instead, technology should imitate natural movements and processes, leading to harmony rather than friction.

By the end of the 1920s, Schauberger was promoting sustainable and environmentally sound methods of technology. He offered ways to combat erosion and water pollution by copying the way water flows naturally. He provided blueprints for ways to generate energy from water. He suggested agricultural practices that increased the fertility of the soil without the need for adding chemical fertilizer – practices based on a sound understanding of the structure, mineral content and polarity of the soil, and the way it reacts with water. He believed that most technological problems could be solved by relying on the principles of cycloid spiral motion, the basic movement underlying all growth in nature. He also believed that, if we treated water right, peace and prosperity would follow.

At various times in his life Schauberger was courted by Hitler, and by Russian and American governments. Tragically, although his work offered the potential to solve many environmental, industrial and technological problems, it also posed a threat to many accepted and lucrative practices. Schauberger died impoverished and alone, his work destroyed by the authorities.

Never forget that what is true of the way water reacts in the natural environment is also true for the way water reacts in our bodies. If we treat our bodies as machines they will suffer stress and ill health; if we revere them, listen to their needs, and nourish them with appropriate rest and the best food and water we can provide, they will respond in kind. Whatever we do to water it reflects back to us; whatever we do to our bodies also reflects back on us.

# Water's Spiral Rhythms

Most of us learn from an early age not to drink from a still pool, but only from a running stream, knowing that the running water was clean and healthy but still water could be full of germs. Still water is dead water because the life of every living thing depends on movement and rhythm – the heartbeat, the circulation of blood, the currents of ocean water, waves pulsating from ocean to shore, the tides. Anything that interferes with these rhythms interferes with the processes of life.

Wherever it occurs, water will always try to form a sphere. In a moving current in a river, this sphere will turn into a spiralling movement because gravity is pulling the currents onward toward the sea. When water falls as rain, or is poured slowly from a jug, it separates into drops with spiral movements in each, forming a perfect sphere before breaking.

Look carefully at a water drop, or a bubble, and you will see the same kind of circulations inside the watery sphere that you see in the steam above tea, or the movement in rivers – eddies, confluences, swirls, spirals and separations of flow in layer upon layer. And these are the same layers in the hydrological cycle: swirls of moist air above the land, swirls of moving currents in the Earth's water, swirls of water transpiring from plant and animal life. Every single drop of water is a microcosm of the universe, containing the same rhythms and patterns – just on a different scale.

None of nature's creative processes involves straight lines, and any water suffers from being directed into straight channels. A naturally flowing stream never flows straight ahead, but meanders in a particular rhythm, and in this meandering flow the water is never still, but moves in layers that continually flow past each other at varying speeds. Streams contain separate spiralling currents revolving to their own rhythm in swirling layers. If you look at the steam over a cup of tea you can see the way the vapour rises in twisting and turning veils – this is a similarly layered movement.

Water is the element uniting all other elements, the link between living things, the essence of all circulatory systems, and its nature is to return to the whole, to a sphere. If something can't follow its natural course it will be frustrated and will lose energy. When water is directed along straight channels and through miles of straight piping, it loses some important qualities. There are obviously huge benefits from access to piped water, as well as some less positive angles. Once we understand the natural processes by which living water sustains itself, we can look forward to ensuring that healthy and energetic water is available for everyone.

> Blood is pumped through the heart in a figure of eight and flows through the body in spiralling currents along the constantly pulsing channels of arteries and veins.

# Spirals of Energy

Spirals of energy lie at the heart of nature: they allow sap to rise, drops to form, shells to emerge and threads of DNA to open and close in response to information transmitted by moving cellular water. Healthy water naturally forms spirals as it flows.

On the surface of a moving stream water flows in a spiralling rhythm from the inside of a bend to the outside, where it turns downward and returns along the bed of the watercourse to the inner bank. Here it rises to the surface again. In every stream the water at the edges flows slower than the water in the centre, so faster layers are continually flowing past slower layers. This creates a labyrinth of extensive surfaces, or boundaries, within the water – these develop and change as layer after layer flow past each other at differing speeds.

Theodor Schwenk, a German teacher and follower of Rudolf Steiner, spent most of his life working with water. He came to think of water as nature's 'sense organ' because of the way it responds to the slightest change in its surroundings.

As one spiralling layer of liquid water meets a different kind of layer, they roll in upon each other to form a series of waves. Such meetings occur when one layer is heavier than another (such as the difference between salt water and river water, muddy water and clear water); or between layers of a different temperature (water at the edge of a river and water in the centre); or between layers flowing at a different rate; or if the water encounters a physical obstruction (a stick or a boulder).

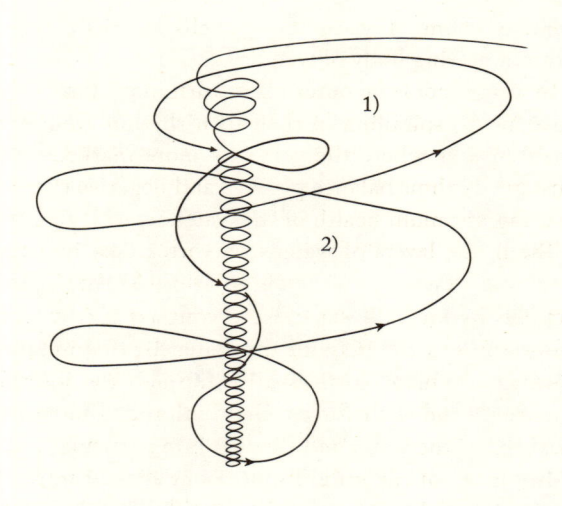

1) Temperatures at the centre of vortices are cooler than those on outer layers: cool water is the most lively and receptive to physical and energetic information.
2) At the core of a vortex the water rotates very rapidly, causing high levels of ionization. This means that the water is able continually to create new combinations and re-combinations of the various elements it carries, enhancing the generation of electromagnetic energies.

Air is trapped in the hollow between the inner surfaces as the wave curls, and is then released into the water as the wave unfurls. These waves produce vortices, vertical spirals of energy within the water that have their own pulsating rhythm. The layers of a vortex turn at different speeds, and the whole form pulsates rhythmically up and down. Water senses every change in its situation, and tries to incorporate any changes in circumstance by momentarily altering its rhythm to accommodate the new situation. A vortex is a typical rhythm within the overall rhythm of water flow, a characteristic feature within a flowing body of water.

In a whirlpool the outer layers turn more slowly than those inside, spiralling in their own rhythm around the inner surfaces where the vortex is more marked. These constant rhythms balance positive and negative energies, ensuring optimum health of the water.

The inside layers of water in a vortex flow in a tight spiral much faster than the outer layers. As the different surfaces rhythmically expand and contract they transmit information to the mass of the water. In the middle of flowing movement vortices arise through the interplay of currents and their forces. The rhythm and movement take hold of the water and shape it into particular forms.

The speed of the water's movement at the centre of a vortex is crucial to its energy. The rapid spiralling motion of the water affects everything it touches. It opens the water to act and be acted upon by outside elements, allowing the water to adapt and change at speed, breaking and re-forming molecular bonds with ease, constantly creating and recreating new combinations of the many elements it carries. As the bonds break and re-form, there is a constant interchange and exchange of electrons; this increases the electromagnetic energy of the water so that it can then be an active carrier of vibrations and physical information.

> *'Observe how the movement of water resembles that of hair, which has two movements, one of which stems from the weight of the hair, and the other from its waves and curls. In the same way, water has its turbulent curls, a part of which follow the force of the main current, the other obeying the rules of incident and reflection.'*
> Leonardo da Vinci noted in 1492

---

## HEALING AT HOME WITH SPIRALLING WATER

- Add minerals to your bath water in spiralling movements to clear it of pollutants.
- Add lavender oil to your bath in a vortex movement, to establish emotional balance.
- Install a simple water feature in your home that incorporates flowing water. The rhythmic qualities will heal your spirit and your environment.
- For garden health follow biodynamic practice and stir vortices into your gardening water (see page 96).

# Spiral Energy

Spiral movement is constant in the development of all living things and systems, whether in the natural flow of water, blood or sap, in the nebulae of space, or even in the movement of the planetary system.

Vortexian energy comes from the effect of the spiralling motion on matter. Such energy depends on the movement of the water, combined with the way this energy reacts with whatever is containing, propelling or directing the water. For example, if you were to direct water along a spiralling wooden channel, the spiralling movement itself provides the water with energy. The fully energized water will react with the density and molecular structure of the wooden channel as the water molecules strike against it.

Energy comes from the friction arising from the water and the wood, from the speed and energy of the one and the density and structure of the other. The release of energy depends not only on the rate of flow of the water, but also on its temperature. Warm still water contains no energy.

# Cooling Waters

Movement is one vital factor affecting the energy of water, temperature another. Water's thermal inertia means that its temperature is not easily affected by minor temperature changes in its surroundings. But small changes in water temperature can lead to dramatic changes in the way water behaves. Just as a change of 0.1°C (1°F) can affect the way that our bodies behave, and can indicate the onset of sickness, so a small difference in the temperature of water can have important consequences on its health and vitality. Water is, literally and figuratively, the Earth's blood.

Water moves quickest when it is cool and is densest before it begins to freeze. It is constantly changing, so water that flows through a warm and sunlit valley will change as it flows into a dark wooded glade. The most energetic water comes from cool and shady sources. Temperatures at the centre of a vortex are cooler than those on outer layers: cool water is the most lively and receptive to collecting and transmitting physical and energetic information to the mass around it.

We can't separate the physical and subtle characteristics of water. They are all integral to its life force. Water is affected by physical information. The movement and rhythm that give water its energy at certain temperatures allow it to carry information in the form of physical matter and vibrations from physical and magnetic realms. All are relevant, all interrelated. When we understand the nature of water we realize the vital importance of good-quality drinking water that can receive and transmit physical and energetic information to your body, thus maintaining health.

Watch trout or salmon leaping up a waterfall on the way to their spawning grounds upstream. They seem to glide up the full height of the waterfall, then swim effortlessly upstream. They can only make these leaps when water is at its peak of energy, cold and full of movement. As a fish dances in the water under the waterfall, at the coldest part of the river, it seems to pick up energy from the vortices in the flowing water and the way these react against its body shape. Then it will leap upward, emerging out of a vortex of water to glide upward without visible effort. Temperature is crucial. At 4°C (39°F) water is densest and at its optimum vitality. Fish can't follow their usual patterns if the water is warmed by even 0.1 degree. The shape of the fish is also crucial – a sort of elongated egg shape that allows the water to flow along and around it in a spiral – thereby receiving maximum energy from the water.

Water that is overexposed to the sun's rays can become tired and lazy – just like us. Cool water at night can become fresh and lively and is able to support and move objects of much greater density with ease. Water is a living creature, responding to, and acting on, its environment.

# Healthy Water Sources

Healthy water needs a diversity of mineral salts and trace elements, both to keep the water healthy and to nurture the human body through their chemical and electrical composition. Drinking water that lacks minerals can cause health problems; drinking water overloaded with the wrong minerals can also cause problems. We all need access to a supply of balanced healthy water.

Water quality depends on numerous factors, particularly at what stage of its cycle we use it. Most useable water can be categorized according to six types: distilled water, rainwater, juvenile water, surface water, groundwater and true spring water. Seawater, which makes up 97 per cent of all the water on Earth, cannot readily be used for domestic supplies (although desalination plants can help some countries stricken by lack of water).

> The receptivity of water – its ability to receive, store and transmit information – depends on rhythm, movement and temperature. So it is not just the information that is added to water that changes the water, but the state of the water when it receives the information. Vital water with a healthy structure is receptive; water with a poor structure is not.

Distilled water is the scientists' 'pure water'. It is not found anywhere in nature, and it contains no dissolved material. Distilled water is the basis for homeopathic and other vibrational remedies as it is blank water, with nothing added. Its purity means that it grasps at everything within reach, seeking to absorb minerals and nutrients wherever it finds them. It is not a good idea to drink quantities of distilled water, because it can leach vital nutrients, trace minerals and elements from you. It can occasionally be used for its short- term purgative effects to remove excessive toxins from the body.

Distilled water is sometimes classified as 'immature' water, lacking those subtle energies that convey and impart health: it has no history, no memory of glacial waters or rocky pools, fast flowing currents or meandering channels, earth minerals, sunshine, or shady places, or any other influences that affect the health of water.

Rainwater is another immature water, unsuitable for drinking long term, although it is marginally preferable to distilled water as it contains some minerals due to its absorption of atmospheric gases and particles. Rainwater can fall to the ground without life, but it then trickles down in spiralling motions around rocks beneath the ground where it gradually meets a rising temperature and begins at some point to percolate upward, always spiralling, gathering mineral ions and life force until it meets light. Rainwater can sometimes also be rather toxic: we have all heard of acid rain, when water collects noxious chemical pollutants and distributes them as it rains, but its lack of life and living energy is more crucial.

Juvenile water emerges from deep in the ground, perhaps in the form of geysers. It has not matured on its passage through the ground, and lacks sufficient minerals to provide drinking water of good quality.

> If you drink much rainwater (for example, if there is no other source of drinking water apart from melted snow, which is frozen rainwater), be careful to supplement your diet. Rainwater lacks potassium and iodine mineral salts; drinking it long term can lead to severe mineral deficiencies, particularly thyroid problems.

Surface water runs off the land to be stored in dams and reservoirs; it contains some minerals and salts accumulated from contact with the soil or from the atmosphere. It is often rather poor in quality, partly because water quality deteriorates if it is stored in still pools, which are exposed to heavy atmospheric oxygenation, and if exposed to the sun. Surface water is also particularly badly affected by deforestation and excessive urban development, both of which change the patterns and rhythms of watercourses – water runs off compacted and deforested surfaces without following its preferred gentle flow patterns which allow it to collect and store physical and other informatio n as it flows. Most piped drinking water comes from surface water or groundwater. Groundwater is usually better to drink, as it comes from deep in the ground, making its way to the surface via various underground passages and streams, maturing as it travels, and containing many trace elements. Some groundwater nowadays may be contaminated by pollution and chemical run-off.

Spring water should be the best quality water, particularly if it gushes out from mountain springs. The best spring water is a shimmering, vibrant-bluish colour; it is very high in dissolved minerals, and is delicious and health-giving. If you get your water from a spring, ensure it is tested regularly to monitor pollution from modern farming practices.

Fossil water is ancient water that can be found in some places heavily protected by layers of clay and rocks. Although fossil water is not moving water it is highly energized. Research suggests fossil water has genuine medical potential. Because it is not part of the water cycle, and therefore not renewable, and is only found in isolated pockets, we must ensure it is never overexploited.

# Bringing Energy to Our Water

Most mains water is groundwater. A proportion is re-treated domestic water which has not been part of the hydrological cycle for some time but has instead been consumed, flushed away into drains, collected, treated, consumed and so on. Even if we may not all have access to the best quality healthy water there are numerous ways we can try and improve the quality of our own water at home. We can either choose point-of-use treatment systems, or turn our understanding of the nature of water to other technologies that can heal, energize and structure our domestic water. A few possible methods are mentioned here. There are many others, ranging from the technically sophisticated to those based on ideas of resonance.

## Detoxifying Water

If you are ill, stressed or just generally out of sorts, one recommended path is to follow a detoxifing regime, then rebalance your body with healthy food, exercise and, of course, by drinking healthy water. The same principle can be followed if water is unhealthy. Several methods of restructuring water begin by completely destructuring the water, then restructuring it with relevant information added. Before it can be used, this restructured water needs to have its life force added through movement (stirring and counter-stirring, or passing through a vortex or flowforms), via magnetism or other energizing techniques.

# Magnetized Water

Magnetic healing has ancient origins. The Indian *Vedas* refer to healing with lodestones, which are magnetite. Plato tells a story of Magnes, a shepherd boy who placed his staff on a rock but had to use great strength to remove it. The boy took tiny pieces from this magnet rock and placed them on his sandals to give him more energy and stamina.

During the last 20 years, scientists from many disciplines, and even alchemists, have experimented with magnets and water. Magnetism can have far-reaching positive effects as it changes the structure of water, and increases its ability to support chemical reactions. If you treat water with negative magnetism then particles of metal, which are positive, will be strongly drawn towards the water. Hans Grander is an Austrian naturalist who has developed a technique to restore natural energy to water by re- imprinting natural biomagnetic vibrations on water (through implosion). The idea is to circulate de-energized water around an enclosed capsule of revitalized water. Various devices have been developed that can be fitted to the domestic water supply. Information from the revitalized water is thus passed to the water running through the pipes, restructuring the water into the revitalized form. Grander water, as it is called, is said substantially to improve the health of people, plants and animals.

> One simple, economic and effective way of cleansing and restructuring water is to put water in a glass bottle and stand it on negative magnets for an hour or two; this pulls pollutants down and dissolves them.

# Flowforms and Further Ideas

Healthy water is moving water that is open to vibrations, and chemical and electrochemical reactions. Just as healthy water naturally forms vortices as it flows in layers of spirals, so dead or sluggish water can pick up energy if we allow it to follow rhythmical patterns of movement. Flowforms are stone or ceramic forms that can be inserted into a water course to oxygenate and energize the water as it is directed through their specially shaped channels. British sculptor John Wilkes created these shapes when working with Theodor Schwenk in the 1970s.

Flowforms mirror water's movement in its natural flow pattern in a stream or river, and also mirror the pattern of blood flowing round the heart. Water courses in a rhythmic double figure of eight through the forms, and as it flows double vortices occur along its path. These energize the water, allowing it to fulfil its potential.

Flowforms are most commonly used in conjunction with aquatic plant sewage treatment systems such as reed beds, where they help to oxygenate the water efficiently and allow bacteria to do their job in cleaning the water. But they are not only useful tools for specific applications, they enhance any garden, and designs are available as interior water sculptures. The gentle rhythmic flow has a peaceful, calming and grounding effect.

Some commercial methods for restructuring water take the electric potential of water as a starting point and use electrolysis combined with filtration to restructure the water. This re-forms water to provide it with a large mass of electrons that are then available to be donated to active oxygen in the body to block the oxidation of normal cells. Water structured in this way is claimed to reduce pain, increase digestive efficiency and slow the ageing process.

Many spas ensure the cleanliness of their water by solarizing it with UV light, or by adding oxygen to the water in the form of ozone – $O_3$. Techniques that heal water are usually also relevant in healing the body. Ozone has long been known to help respiratory problems and modern research now suggests that ozone therapy can benefit individuals suffering from problems with their immune systems: it oxygenates the blood and may stimulate production of white blood cells.

---

### ALL OF US CAN USE THE FOLLOWING SIMPLE HEALING TECHNIQUES AT HOME:

- If you can't afford flowforms, try creating vortices in the water before using it (see pages 84-85, 129).
- If you can't afford commercial devices, try resonance ideas such as directing the energy of crystals at your water, or opening it up with colour (see page 76-77).
- Try healing your water through the power of thought alone, transferring your healing energy to the water.
- You could solarize water (see page 72), or treat it with pyramids, music, prayer, homeopathy or other vibrational therapies.

# Biodynamics

Food that has been biodynamically grown, where water and soil care is of utmost importance, is extremely good to eat. This increasingly popular method of agriculture or horticulture was developed by the educationalist and philosopher Rudolf Steiner. It takes account of cosmic as well as physical factors when planting and harvesting – planting according to the movement of the moon, the planets and the time of day as well as the seasons, so that plants can take advantage of all the most positive conditions for growth.

One rather bizarre example of water practice and good husbandry comes from old-fashioned farmers in Austria. Viktor Schauberger tells the story of meeting an old farmer whose crops always yielded much better than his neighbours. The old man stood in front of a large wooden barrel full of water, stirring it and occasionally throwing in a handful of loamy soil. He stirred first to the right, then to the left, and as he stirred he sang quite loudly at the water. Toward evening, clay would be stirred into cooling water with a large wooden spoon. When stirred toward the right the mixing would be accompanied by ascending notes, and when stirred toward the left, by descending tones.

This practice, known as *tonsingen*, can be explained in terms of good husbandry. The clay loam would be stimulated by a variety of vocal sounds as a farmer would learn, through experience, at what pitch his voice would resonate with the contents and/or the shape of the barrel.

As the fermentation process took place in cooling water in darkening light, the $CO_2$ breathed out by the ferment would become bound in the water. The singing and stirring encouraged the charges of metals such as aluminium to break from the clay and become finely distributed in colloidal form throughout the water. The water therefore became neutrally charged. It was sprinkled over newly sown fields where it eventually evaporated, leaving very fine crystals. These crystals carried a negative charge and attracted rays from all directions, keeping the soil moist, cool and healthy and in ideal condition to support the crops.

One biodynamic treatment, '501', is made from finely ground silica, which is placed inside the horn of a cow and buried in the earth over the summer. The silica picks up increased energy from its period in the earth in its own spiral container. A small quantity of this powder is put into a barrel of clean spring water and stirred vigorously clockwise and then counterclockwise for about an hour. This water is then sprayed or sprinkled onto compost, pastures and crops. The stirring sets up vortices in the water, encouraging distribution of the minerals. The water is highly energized and able to receive and transmit energetic information from the silica.

# Drinking Water Troublemakers

- *Aluminium* occurs in water in some areas from acid rain, or it can be added to clear cloudy water. Some hot water tanks have aluminium anode rods. Aluminium sulphate is added at water treatment works as part of the cleansing process. Excessive aluminium intake contributes to premature ageing and Alzheimer's disease. But remember, we all need traces of aluminium in our bodies – it is powerful in combating radiation.
- *Nitrates* leach into the water as residues from fertilizers. Many can cause cancer.
- *Herbicides and pesticides* also leach into the water from farming practices; little is known about many of the health risks, but herbicides such as atrazine, now banned in Europe but often found in US rural drinking water supplies, may even cause birth defects.
- *Heavy metals* can be found in water – pipes in many older houses are made of lead. Excess arsenic in water causes illness, defects and death.
- *Chlorine* is added to sterilize drinking water against bacteria, but this may carry its own risks. Chlorine affects the metabolism of fat, and hormonal activity; it inhibits the actions of certain enzymes and may inhibit recovery in some illnesses. Some people are allergic to it. Chlorine has been associated with heart disease, and it kills the bacteria in the intestines, which have multiple functions necessary for good health. If you drink unfiltered tap water, fill a jug and let it stand while some of the chlorine evaporates. Then add a teaspoonful of vitamin C to each litre of water to combat effects of chlorine.
- *Trihalomethanes* (chloro-organics) occur when chlorine reacts with untreated sewage. THMs are thought to be carcinogenic.

Water is good for you, so don't be scared to drink it – if your supply isn't satisfactory, do something about it! We need to know what is in our tap water, so we can make use of appropriate systems to cleanse and energize it. Also, we can adjust our diets if necessary to take account of the pollutants in our water, but remember that it will do us little or no harm to ingest minute quantities of most metals.

Ask your water provider for an analysis of the health of your water.

- There may be a spring or well near you, check it out and get the water analyzed.
- Find out about filtration systems that will suit your home, and your pocket.
- Fluoride is commonly added to water as a 'public service' to prevent tooth decay in children. Controversy has surrounded this practice as large amounts of fluoride can weaken the immune system and may cause heart disease, genetic damage and cancers. Naturally occurring fluoride is one of the most toxic substances present in the Earth's crust, only marginally less so than arsenic. Eating organically grown wholewheat bread can help counteract the effects of fluoride.

Fluoride inhibits a wide range of enzymes in the body – by distorting their molecular structure or combining with essential trace elements. Fluoride can weaken the structure of bones, and it appears to inhibit the functions of white blood cells, therefore affecting the immune system. It is known to have bad effects on kidneys, heart, thyroid gland, eyes and skin, as well as impairing brain function. Fluoride may even seriously inhibit the body's natural DNA repair system.

# Filtering Your Water for Health

The first step in any water treatment programme is finding out what your water contains. Your local water supplier should be able to give you a copy of the water analysis of your supply. These results will tell you about the water when it leaves the treatment centre, but don't forget that water is the most effective solvent and it will pick up chemical information from the pipes on its way to your home, and is also subject to electromagnetic information from domestic and environmental sources. There are a variety of home treatments that can help the health of your water.

- Carbon filters are the simplest, but they will not treat bacteria or viruses. Carbon removes impurities through mechanical filtration and adsorption. The enormous surface area of the carbon functions as an electrochemical sieve, removing various contaminants (e.g. chlorine), unpleasant tastes and smells and some human-made chemicals. Solid block carbon filters may be more effective than granulated activated ones, where bacteria can multiply within the carbon bed. If you don't use a carbon filter for more than a day or two, run the water through it for at least 30 seconds before use.
- Distillers boil your drinking water, allow the steam to condense, and leave behind unwanted substances. Distillation removes harmful contaminants from water, but drinking distilled water is only valid as a short-term therapeutic detoxifying measure.
- Ceramic filters consist of a ceramic container riddled with holes that allow water through but trap bacteria. Some contain an ion-exchange mechanism that also traps some metals and helps soften the water. They are particularly suited where there are bacterial problems.

- Reverse Osmosis (RO) filters are by far the most effective, widely used in therapeutic establishments to ensure clean pure water. They force water through a semipermeable membrane. The membrane allows the water to pass while rejecting most of the contaminants in it. These are then flushed to the drain and the purified water collected in a tank. Reverse Osmosis filters can remove almost all contaminants. They are effective in reduction of heavy metals such as lead, mercury and arsenic, and reducing nitrates, herbicides and pesticides; they can even reject bacteria and viruses.
- Table-top carbon filters are cheap, convenient and recommended as a first foray into filtration. Make sure you change the filter at least as often as the manufacturer recommends as they can become ineffective or even dangerous as bacteria that have collected or farmed on the carbon seep into filtered water. Reverse Osmosis systems produce the cleanest, purest water of any filtration system. The water is believed to be effective even against excessive radiation. The best RO systems can change your domestic water into pure water of medical quality. Most units are rather bulky and use significant quantities of water to keep the membrane washed clean. In Israel, every member of the army carries a portable RO filter because of their concern about the effects of possible nuclear contamination.

# Bottled Water Questions

Increasing amounts of money are spent each year on bottled water, assuming that it is healthier than tap water. But this is not necessarily true. Some 2,000 years ago – long before the problems with water quality that we have now – the Romans bathed in the springs where world-famous Perrier mineral water comes from in

France, but they wouldn't have considered drinking it! Bottled water is useful when you're travelling, and is a healthier alternative to other soft drinks, but it is not necessarily healthy in itself.

- Natural mineral waters must come from underground sources and be free from harmful bacteria. There is no minimum level for the waters' mineral contents. Some filtering is permitted, and the water may be carbonated to make it sparkling, but it can't be disinfected and nothing else can be added or taken out of it.
- 'Spring' or 'table' water simply has to conform to the same quality standards as tap water. Waters with these names can come from any source, from natural springs through to mains water. Suppliers can transport and blend water from different sources, filter it and treat it. So you may be buying mains tap water that has been filtered, probably treated with ultraviolet light to kill bacteria, and possibly carbonated. This water probably costs 500 times the cost of the original tap water!
- Stop and think before you buy bottled water: it may have been standing in its bottle in a warm warehouse for many months, or possibly even years. Would you drink tap water that you had poured weeks, or even days, previously?
- Some still waters contain high levels of bacteria, particularly as they may be bottled and stored, unrefrigerated, for up to 18 months before they reach the customer.
- Sparkling bottled waters tend to have the lowest levels of bacteria, because the carbon dioxide inhibits their growth.
- Storage time and conditions affect the quality of bottled water. Even if the water is bottled at source, and is rich in particular minerals, the energy in the water will change if it stands around, and the water will assimilate some information from the plastic packaging. Some unhealthy chemicals

can leach from plastics and packaging. For example, oestrogens have been reported to leach out of plastic through compounds known as alkylphenols or biphenolics. Even if migration levels are low, there could be some risk, as increased oestrogen levels can cause cancers and genetic problems. Bottled water is most useful for emergency supplies, when travelling or staying away from home. Once you have opened a bottle of water drink it all at once rather than storing it. Bacteria multiply as soon as the water is opened.

# Drinking Water for Health

When feeling unwell, run-down or simply out of sorts we tend to look for complicated solutions, but the answer may be incredibly simple – just drink more water! Water may be almost the only medicine most of us will ever need.

One of the major causes of ill health is dehydration. We are all watery creatures, we need water. If we don't drink enough we can harm our bodies and our brains – body processes can't function so efficiently and we can suffer stress and depression as well as a wide range of physical ailments. When we starve body cells of water they start to complain, and many adverse reactions set in. A body can survive without food for several days, even weeks, but without water, no one can last more than a few days.

Every body needs 2 litres (3.5 pints) of water a day. If possible drink filtered water; other drinks do not have the same effect. When you take in water as part of a drink, or in food, your body separates the solids from the water; this normal body mechanism is complicated if the drink or food is diuretic since more water will be eliminated from your body than added. Tea, coffee and alcohol are all diuretics that drain water and minerals from your body rather than replacing fluid.

It is best to drink when you get up, before you go to sleep, half an hour before eating and between meals. Don't drink water with your meal, since water's role is to transmit information to the body and to flush out toxins. If you drink at the same time as eating, this interferes with digestive processes and may lead to indigestion, acidity or nausea. But a drink before eating helps your body use the water to lubricate the whole system, making digestion easier. You will also be less likely to suffer from excess cholesterol formation.

As you increase water intake, make sure you take care of your diet because salts and minerals can initially be drained out of your body. Remember to keep up your vitamin and mineral intake. If in doubt, eat carrots – beta carotene stimulates take up of vitamin A, and is essential for liver metabolism. A healthy liver is central to good health because it is the body's main filtering and fat-burning organ.

# Here Are 10 Steps to Health Through Water:

- Start the day with a glass of water. This flushes out your kidneys and detoxifies your system for the day ahead. Drink at least eight glasses of water daily. If possible, drink filtered water.
- Listen to your body. Don't eat if you are not hungry. Drink a glass of water instead.
- Learn to recognize what your body is telling you – if you suffer from headaches, stomach pains, fatigue or other common ailments, try drinking a glass of water before you reach for a proprietary medicine.
- Drink a glass of water half an hour before you eat. Don't drink with your meal. You absorb water through your food, so whenever possible, eat organically grown food – its water content will not be contaminated.
- Drink a glass of water if you feel stressed or anxious. This will help to keep your body fluids flowing smoothly, and can help calm you down.
- Eat sensibly. You don't need to count calories, carbohydrates or fats but be aware of good intestinal hygiene: your liver needs to filter out and destroy any bacteria or viruses in the food and drinks that you take in, so drink plenty of fresh water and avoid overdosing on processed foods.
- Don't drink too much tea, coffee or alcohol. On occasions when you do overindulge in any of them, drink more water to compensate.
- End the day with a glass of water.

# Drinking Water for Common Ailments

Historically, the water in the body has been assumed to act primarily as a solvent, a space filler, and a means of transport, with the substances dissolved in the body's fluids considered the major players. These are thought to be the regulators of the body's activities, responsible for specific reactions or diseases. However, Dr Fereydoon Batmanghelidj, founder of the Foundation for the Simple in Medicine, asserts that it is the water itself that regulates the body's health or otherwise. The following recommendations are based upon his work.

## Arthritis

- Treat pain initially as a sign of dehydration and combat it with increasing your intake of drinking water.
- If you don't drink enough your blood can become too viscous, cells can complain, blood vessels may become constricted, joints dry and dehydrated. In arthritis, cartilage surfaces dry up and movement becomes painful as joints become dehydrated. This can eventually cause severe damage, and although this damage will itself trigger mechanisms for repair, this process typically produces a disfigurement of the joints. To avoid this, as soon as you begin to suffer from any kind of joint pain, it is a good precaution to recognize the pain as a sign of local dehydration and compensate by drinking more water.
- Millions of people suffer from some form of arthritis, with a disturbing upswing in juvenile cases. It is possible that many people's suffering could be alleviated or even cured simply by following a responsible drinking water regime. Such a routine can certainly do no harm, and may be extremely beneficial.

# Asthma

- Maintain a healthy drinking regime, alongside proper medical supervision. Asthma and allergies are exacerbated by many of the toxins in modern life – in the air we breathe, the food we eat and even through drinking bad water. They occur when the body's regulating neurotransmitter systems (histamine and its agents) become excessively active. Asthmatics have increased levels of histamine in their lungs, and if concentrated blood reaches the lungs they will produce more histamine. Exaggerated release of histamine leads to bronchial constriction. Conventional medicine suggests asthma and allergies should be treated with suppressant drugs; Batmanghelidj recommends that rises in histamine levels should instead be treated by increasing the body's level of water intake so that histamine's purpose can be satisfied rather than blocked.
- A good basic practice is to gradually increase drinking water intake and maintain at least eight glasses (2 litres/3.5 pints) of quality water a day.
- Asthmatics should not drink more than one small glass of orange juice daily as the potassium content of orange juice is high and this can increase histamine production.

## Backache

- Drinking sufficient water and following regular exercise ensures joints and discs will keep lubricated and function properly.
- Lower-back pain is a common complaint. Spinal joints and their discs are dependent on the different hydraulic properties of water stored in the disc core and the cartilages. In the joints of vertebrae, water is not only the lubricant for the contact surfaces, it is held in the disc core within the intervertebral space and supports the compression weight of the upper half of the body; 75 per cent of the weight of the upper part of the body is supported by the water volume stored in the disc core. Intermittent vacuums created by the movement of the joints promote circulation of the water. As a sensible start to any regime, take care to avoid dehydration.

## Constipation, IBS, Urinary Problems

- Eat fresh food and drink more water.
- Prevent problems by drinking sufficient water for the body to process its waste products. Constipation causes pain, stress and sometimes long-term problems of the colon and bowel. Drink a minimum eight glasses of water a day and take regular exercise. Increasing water intake will often relieve Irritable Bowel Syndrome (IBS).

# Drink Water, be Slim

- Forget diets, counting calories or craving chocolates – you can lose weight simply by drinking a glass or two of water before a meal! Sensations of thirst and hunger stem from low energy levels. People often fail to differentiate between the two but if you drink before you eat the sensations will be separated and you shouldn't overeat.
- Whenever you feel like reaching for a snack, reach for a glass of water instead.
- Eat regular meals, drinking a glass of water half an hour before each one. Weight problems should be eased, and your energy levels will be boosted.

# Headaches

- Before you search out any proprietary medicine, reach for a glass of water. Headaches are usually a sign of some kind of stress, which may be relieved through simply drinking water.

# Heartburn, Indigestion

- Drink a glass of water slowly, and follow this with another glass half an hour later, and a third half an hour after that.
- Some stomach pains may indicate severe problems, including perforated ulcers, but this is comparatively rare. If pain is prolonged and persistent, you should, consult a doctor, but most dyspeptic pain – gastritis, duodenitis, and heartburn – is associated with chronic dehydration. It typically occurs when there is insufficient water to keep the mucus lining of the stomach hydrated, allowing acids to get through from the stomach to its sensitive outer layers, causing pain. Rather than treating the condition with antacids – which may have their own unwanted effects – what the body needs is water.

# High Blood Pressure, Hypertension and Heart Problems

- These may develop because blood cannot flow freely through the system. If you don't drink enough water your blood can become thick and sluggish, causing constrictions and high blood pressure. Increased water intake can be particularly helpful.

## High Levels of Cholesterol

- To decrease cholesterol levels, drink between 8 and 12 glasses of water a day. It is even more important to watch what you drink rather than being excessively rigid about what you eat.
- High cholesterol levels do not occur as a direct result of eating fatty foods, but because of a lack of water in the diet. This leads to cholesterol building up on the inside of the arteries rather than getting dispersed. Cholesterol is a natural substance that covers cell membranes to make the cell wall impervious to the passage of water. It is part of the cell survival system. It is manufactured to protect living cells against dehydration.
- If you sit down to eat without drinking a glass of water previously, the process of food digestion will take its toll on the cells of the body. One result will be the formation of excessive cholesterol. The integrity of every cell membrane depends on the presence of available water.

## Menstrual Problems

- If you suffer from pre-menstrual tension or stomach cramps at period time, insufficient water in your diet may be the major contributory factor. At times of hormonal changes, your body needs additional support, and extra water intake helps biological processes to function smoothly.
- As your body loses fluids, it needs to manufacture more, and it needs enough water intake to be able to do this. If you instigate a healthy drinking regime, drinking at least eight glasses of water daily, tension and stomachaches should disappear as your body will have enough fluids to replace those it is losing.

## Morning Sickness

- Increase your water intake gradually.
- Developing babies live in an entirely watery world, and pregnant women need to be particularly sure to keep up their fluid intake. When a mother suffers from morning sickness this not only causes dehydration, it can also be a sign of dehydration. One of the best solutions is to maintain a sensible routine of drinking water, making particularly sure to drink a glass of water first thing in the morning and last thing at night. Unfortunately, some women find that drinking tends to make them feel more sick – in that case, try drinking small amounts often until accustomed to a regular routine.

## Stress

- Stress can also be exacerbated or even caused by lack of water. When the body becomes dehydrated the physiological processes that come into play are similar to those it uses to cope with stress.

# Chapter Four
# Bathing

Bathing is not just about getting clean. Regular bathing is one straightforward way of taking control of your physical, emotional and spiritual health. The simple action of taking a bath can lead you to feel at peace with yourself and with the world around you, calm troubled emotions and lift the spirit. Water lies at the heart of all life's processes, and bathing can unlock and transform many energies vital to life.

## The Birth of Hydrotherapy

Early systems of healthcare understood that water was the basis of all health, the element linking humans, the earth and the heavens. Traditional Chinese medicine saw illness as a sign of disharmony within the whole person, and water was (and is) central to restoring the harmony as it is the carrier and mover of *chi*. Indian Vedic healing similarly describes water as the element linking the individual to the cosmos, believing that the *prana* in water can be transformed and converted into other forms of energy that will invigorate and strengthen the human body.

The ancient Egyptians had an integrated view of the sacred and physical properties of water, they built temples where priests bathed and drank mineral-rich waters, and water treatments became established as effective therapy for many physical and spiritual conditions. Highly evolved bathing practices also existed in Sumerian and Babylonian societies.

The Greek Hippocrates (died c.430-c.370BC) advocated regular bathing to strengthen the constitution as well as water applications for specific complaints. Galen, the Roman who formalized many Hippocratic theories, also supported the idea of bathing for particular treatments and as a means of rising to a higher level of health. Bathing was an integral part of Roman society, supporting beliefs in health, fitness and moral strength.

Hindus believe that water has a soul, so bathing is often seen as a spiritual ceremony, a way of restoring harmony of body, mind and spirit and connecting with the divine realm.

Every town in the Arab world boasted *hammams*, or Turkish baths, as early as the 4th century, but bathing practices in Europe disappeared with the Roman Empire and only re-emerged in the 14th century. Bathing was primarily a secular and functional activity, rather than the sensual and spiritual affair of the early bathers.

# The Work of Priessnitz and Kneipp

By the end of the 18th century, wherever there were natural springs or sources of mineral-laden waters, water cures were springing up, particularly throughout Northern Europe and the US. One of the most famous was the Water University of Grafenberg, set up in the mountains of Czechoslovakia by a Silesian farmer, Vincenz Priessnitz (1799–1851). Clients ate a simple diet, took regular exercise and massage sessions, and had dozens of water treatments. They drank up to 5 litres (9 pints) of water a day to cleanse the blood and the kidneys, had enemas, and were immersed in hot and cold baths to stimulate the circulation.

Vincenz Priessnitz is often recognized as the father of modern hydrotherapy. He introduced the foundations of naturopathy, a system of healing using nature's elements of air, earth and sunlight with water. Although naturopathy appeared to help bridge the gap between emerging medical practice and traditional healing practices, it lacked the spiritual dimension of early systems.

Sebastian Kneipp (1821–97) further developed Priessnitz's work. He believed that blood obstructions are the main cause of disease, recognizing the importance of maintaining freeflowing circulation (this recalls traditional Chinese or Indian medicine). Although a rather unhealthy young man, Kneipp was convinced the human body could heal itself, given the right conditions, and that water was instrumental in the healing process. He learned ways to use water, with herbal treatments, and his versatile hydrotherapy regime enabled him to live a long and healthy life and practise his calling as a Dominican priest.

Kneipp's treatments generally referred to specific conditions (such as colds, bronchitis, catarrh, muscle pains or circulatory problems) rather than the constitution-based water healing of earlier philosophies. They often involved patients immersing themselves totally in cool or warm baths, or taking 'sitz baths' – sitting baths. Central to his therapies was the observation that water temperature affects the circulation, and bathing in alternate hot and cold water was considered a remedy for a wide range of complaints.

These early hydrotherapies are still popular, and valid, but have a puritanical feel to them – cold baths, cold showers, cold compresses or uncomfortably hot alternatives. Yet water therapy need not be unpleasant! Bathing can be sensual and fulfilling, pleasurable to all the senses, and provide lasting benefits.

# Kneipp Baths

## Treatments for Common Complaints

### COLD BATH TO RESTORE ENERGY

- Fill the bath with cold water (about 15.5°C/60°F) to the depth of about 25 cm (10 in). Lower yourself to sit in the bath, rubbing the water gently over your whole body for about a minute. Then dry yourself vigorously. You will feel invigorated for hours.
- Kneipp also recommended body brushing or pouring cold water over the body as a remedy against tiredness. Body brushing also stimulates the lymphatic system, which helps the body to get rid of waste products.

### HOT AND COLD AGAINST ACHES AND PAINS

- Kneipp recommended sitting in alternating hot and cold baths (sitz baths) but a shower is as effective. Start with a nice warm shower for 2–3 minutes, then lower the temperature to cold for up to 3 minutes, then raise it again to warm, and lower it again. Repeat the process several times.
- Aches and pains will be eased, inflammation will be reduced, and your blood circulation will be boosted so you feel more energetic. This is also a good constitutional remedy.

## CALMING AND DEEP CLEANSING BATH

- Get into a bath half filled with water at body temperature. As you sit in the bath, slowly add cold water and lower the temperature to 15.5°C (60° F) in 15 minutes. Then get out, wrap yourself in a large warm towel and rest for an hour.
- This bath is very relaxing to treat stress and insomnia, and acts as a detoxifying bath to eliminate alcohol, nicotine or caffeine as it encourages the kidneys to work hard. The effects can be even more marked if you drink valerian, thyme or fennel tea before bathing.

## BATHS FOR PELVIC PROBLEMS

- If you suffer from menstrual pain, pelvic inflammation or urinary tract infections, try sitting in a hot 44°C (110°F) bath for 5 minutes, followed by a cool rub down.
- Or make a herbal infusion by steeping equal quantities of juniper berries, rosehips and horsetail in boiling water. Strain and add to hot 40°C (104°F) bath water. It is very soothing to sit or immerse yourself in the water for about 10 minutes.

# Kneipp's Treatment to Reduce Fevers in Children

> Caution: These treatments can be used to make a slightly feverish child more comfortable, but if a child's fever continues for more than 24 hours, or is severe, you should always consult a medical practioner.

## HERBAL BATH

- Prepare an infusion using equal parts of salt, cider vinegar, thyme, chamomile, fennel and lime blossom in boiling water.
- Leave to cool to room temperature, strain the liquid and add to warm 24°C (75°F) water in a shallow bath, making sure the bathroom is warm. Bathe the child for a few minutes, then dry them in warmed towels, dress in warm cotton nightclothes and put to bed.
- If the fever continues, use the same infusion to make several compresses which should be placed on the chest and back of the child, covering the child with large warm towels. Repeat as required until the temperature falls.

## SALT BATH

- Dissolve 2 cups of sea salt in 2 litres (3.5 pints) of cold water. Make sure that the room is warm, then wash a feverish child with this water before dressing them in warm cotton clothing and putting to bed. The minerals in sea salt help detoxification and rebalance the body's energy. Rest is essential after a salty bath, so the body can rebalance gently.

# Skin and Water

Our skin, more than anything, indicates our general health, and it is vital in maintaining and regulating our health. It is in fact the body's largest organ, important for detoxification and vital for absorption – both of which happen through the pores of the skin, the hair follicles, and the sweat glands.

The skin is also a significant organ of communication, telling us what is hot, cold, wet, dry, rough or smooth; it transmits pressure, itching, pain or pleasure. Sensations of touch instantly evoke reflexes. All our senses can be profoundly influenced through immersing our skin in water or surrounding it with aromatic steam.

We all have between 3 to 5 sq m (32 to 54 sq ft) of skin, up to 15 per cent of our body weight. It is covered with a fine but extensive and complex network of nerve endings, which represent the major sensory receptors so signals from the skin communicate directly with our central nervous system. Special receptors in the skin help maintain the body's even temperature.

When the skin is cold, the arteries, veins and capillaries become narrow to reduce the blood flow and conserve body heat. When the skin is hot, blood vessels relax and dilate to allow the blood to reach nearer the surface of the skin in order to shed the body's internal heat through perspiration. Sweat glands pour moisture on to the skin's surface, which produces a cooling effect when it evaporates.

Sweat glands are controlled by the hypothalamus situated in the brain – the hypothalamus controls body temperature via the pituitary gland, which has a vital role in regulating fluid balance. The link between the sweat glands in the skin and the brain causes us to perspire when we are nervous, agitated or emotionally stretched or stressed.

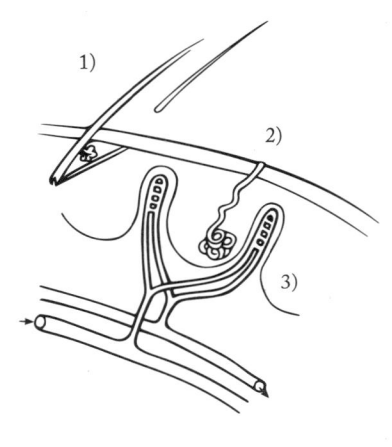

## Cold skin

1) Hairs stand erect
2) Sweat glands that produce no sweat
3) Blood supply to the skin closes off to conserve heat

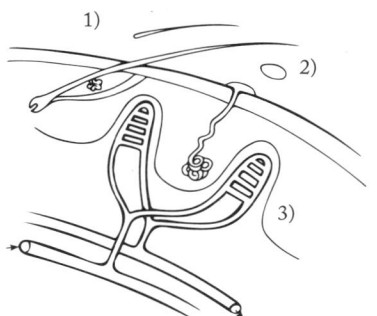

## Hot skin

1) Hairs lie flat
2) Sweat glands that produce sweat
3) Blood supply to the skin opens up to release heat

- *Hot and cold water* send different nerve impulses to the rest of the body. The effect of water on the skin is the key to hydrotherapy.
- *Cold water* is stimulating: it makes surface blood vessels constrict, restricting blood flow, and inhibiting the biochemical reactions that cause inflammation. Cold water also sends blood toward the internal organs, helping them to function better.
- *Hot water* is relaxing: it dilates blood vessels, reducing blood pressure, and increasing blood flow to the skin and muscles,

easing stiffness. Improved circulation boosts the immune system, helps remove waste products from the body, and sends more oxygen and nutrients into the tissues to repair damage.
- *Bathing alternately in hot and cold water* stimulates the hormonal system, reduces circulatory congestion caused by muscle spasm, relieves inflammation and balances blood pressure by narrowing and dilating the blood vessels. It also increases our body's electrical potential, and therefore its ability to receive vibrations or energy, and it stimulates the lymphatic system into activity, which encourages efficient waste-clearance.

# Minerals in Bathing Water

Health problems change over the centuries, even over the decades; we no longer have to worry about many pests and diseases that were once life threatening, but other conditions arise in their stead. One scourge of modern health is environmental pollution. With the best will in the world, we can't avoid pollution, but we can do our best to counteract its effects.

Every day we are exposed to toxins and pathogens, in the air we breathe, the food we eat, the water we drink, the places we work and even the homes we live in. Our environment is contaminated by heavy metals and chemical pollutants from industrial and agricultural processes, through organophosphates, nitrates and organochlorines in our food and water, and by electromagnetic fields penetrating most areas of our existence. All these pollutants contrive to weaken the body's natural defences, but we can overcome the effects of this daily invasion by looking after our health.

Each of us has a unique ability to resist or overcome invading pathogens. This is determined in part by our constitution with its specific strengths and weaknesses – genetic and acquired – and by our lifestyle and environment. Every one of us has a unique metabolism: our ability to detoxify what we do not need, to absorb what is essential to life and to maintain the correct balance between the two.

The simple act of bathing can bolster the constitution, stimulate the metabolism, and combat the effects of pollution. Immersion in mineralized water can change the structure of the water in our bodies, making it able to receive the healing vibrations that are the source of our life force.

# We Need Minerals

Minerals make life possible. They make up our rocks and soils, then feed the waters of the world. They are the main source of nutrition for every human, animal and plant. Our bodies require large amounts of some minerals but only trace amounts of others. Minerals and trace elements are both equally essential to health – we need a few milligrams of zinc just as much as several hundred milligrams of calcium.

Calcium, phosphorus, potassium, sulphur, sodium, chlorine, magnesium, iron and silica are the main minerals our bodies need, along with over 60 trace minerals. These include zinc, copper, boron, cobalt, chromium, bromine, gold, selenium, lithium, molybdenum, manganese, germanium, fluorine and tin. Whereas calcium is so important to our structure that it makes up between 1.5 to 2 per cent of body weight, each trace mineral makes up less than 0.01 per cent.

Organic gardeners know that nutritionally balanced soil will mean healthy plants. A nutritionally balanced body, with all the required minerals in the correct balance, makes us healthy, too. Mineral imbalances lead to ill health and malfunction. We obtain our essential minerals from food and water, but we can't store them in our bodies – they need to be constantly replenished through food and fluids.

Minerals are nutrients that are necessary for the structure and function of cells, and are part of the body's biochemical make up and physiological processes. They act as essential catalysts for the body's manufacture and assimilation of the vitamins, enzymes and amino acids upon which health depends. They are instrumental in biological reactions such as transmission of nerve impulses, digestion, use of nutrients and in hormone production. They maintain the delicate water-acid base in the body that is essential for proper functioning of all mental and physical processes.

If we are exposed to too many toxins and pathogens, our blood will typically become acid. The more acid the blood, the less oxygen it can carry – as well as many other problems, this leads to vulnerability to invading viruses and bacteria, premature ageing and a damaged immune system. Blood acidity also impacts on the acid and alkali balance of the entire system, including saliva, mucus, stomach acid and our skin. So the correct balance of minerals is vital in optimizing the body's defences against pollution, stress and infections.

Environmental pollution means that we have to be more and more careful to ensure that we do have an adequate intake of minerals. While a healthy diet is vital to maintain adequate mineral levels, in today's polluted world bathing can have an equally important role, when we can absorb vital minerals through the skin. Sea bathing, salt-water bathing, spa bathing and mud bathing are particularly relevant.

> Used as a healer by the ancient Greeks, horsetail is a strong astringent, diuretic and tissue healer. It contains some of the most valuable minerals the body needs: silica, potassium, manganese and magnesium.

Just as modern chemical pollutants and electromagnetic disturbances affect the water, changing its structure and diminishing its power to carry or transmit energy, so minerals' electrical energy affects the electromagnetic field of the water and its ability to carry healing vibrations. Bathing in mineralized water is one way of tapping directly into water's vital energies.

The mineral springs at Saratoga Spa State Park in New York State remain a major attraction. Mineral-rich water bubbles up all around the town, where it is bottled for drinking. One of the best- known springs incorporates a breathing port at one side for inhaling carbon dioxide, said to be good for the lungs and sinuses. The gas also carbonates the water and powers geysers that spout water high in the air. Another spring has cloudy greenish and rather smelly water said to have particular diuretic qualities, containing large amounts of lime, sulphur, iron and other minerals. Spa visitors can take relaxing mineral baths floating in the salty sparkling water: studies have shown that some carbon dioxide is absorbed through the skin, where it dilates the blood vessels and aids the circulation.

# Sea Bathing

The sea is the lifeblood of the planet. The sea affects the energy of us all, wherever we live, since its waters carry celestial as well asphysical information from shore to shore. The flow of every drop of liquid on Earth mirrors the tides of the oceans, from the blood in our veins to the water we drink. So our life force is affected directly by the great water masses that cover the Earth.

The sea conjures a vivid picture of health, happiness, energy or tranquillity. The sea means something to all of us, and nothing can beat it as the perfect tonic, rejuvenating body and soul. Fill your lungs with salty air, savour the smells of the sea and the sounds of the waves. Sit on a rock and watch the sea's unceasing movement. Relax your mind, let your thoughts wander. The gifts of the sea will make your energy levels gradually rise. Bathing in seawater can renew flagging energies, heal mental and physical problems, and strengthen mind and body to prevent illness.

# Seashore Energy

Wherever we are – city, farm, mountain or seashore – the air contains roughly the same proportion of oxygen (21 per cent); yet sometimes it feels heavy, and at other times it feels full of energy.

Air quality is partly determined by the movement of electric energy, and the way that movement interacts with the electric and magnetic fields on and around the Earth. This is partly due to weather conditions, as the Earth's electric field is affected by sun, rain, wind and particularly thunder and lightning. The heavy and uncomfortable feeling before a storm is due to the build-up of positive ions in the atmosphere, which will only be broken when rainfall or lightning brings the necessary high negative ions to overcome them. Apart from weather, the chief factor affecting air quality is pollution, where chemical pollutants steal negative ions from the air.

While pollution tends to clog up the air, leaving too many positive ions, the energy in moving water lets loose a mass of negative ions: as the water breaks up into drops, the positive ions remain with the larger drops, and the negative ions fly free with the fine spray. The finer the spray, the more negative ions are produced through the friction of air and water. At the seashore (and in the mountains) the air is always negatively charged, encouraging deep breathing and increased energy levels.

In modern urban surroundings, tracts of land are covered with asphalt and concrete, altering the Earth's electric field, creating electromagnetic stress and preventing the natural release of negative ions into the atmosphere. The pollution in cities produces more and more positive ions, with the result that we may feel tired, irritable or just generally out of sorts.

Serotonin, sometimes known as the 'happy chemical', is a neurotransmitter produced by the brain to control our moods – air that contains too many positive ions is thought to depress the brain's ability to release serotonin, making us prone to an imbalance in our emotions. This may make us over-stressed and anxious, or lethargic and depressed.

Tuberculosis (TB) clinics and sanatoria were historically sited near the sea or in the mountains where the air is rich in negative ions. These ions combat the effects of pollution, boost energy and have an antibacterial effect.

# Seawater Energy

It is thought that all natural elements exist in the sea, fed into it via thousands of rivers and streams. As water evaporates, the minerals are left to form the bitter saline brine we know as seawater. Seawater's antibacterial action clears skin infections from inside and out. Bathing in seawater is a good general health tonic, as it contains vitamins and minerals that are almost identical to those in our own blood's plasma. Blood plasma is responsible for feeding and strengthening all our cells to make sure our bodies' function healthily.

At blood temperature the minerals present in seawater can penetrate into the bloodstream to be absorbed by the cells that need them. They rebalance the whole body. As well as all-important minerals, the sea contains vital proteins and vitamins in algae and seaweed. These elements help to feed and detoxify the body via the skin. The iodine content of seaweed is a powerful natural antiseptic and also has a specific role in balancing the thyroid gland, which controls the body's biochemistry.

Seaweeds and other algae are the oceans' cleansers. Their extraordinary ability to absorb heavy metals and other pollutants means they can keep seawater clean enough to support marine life. Their deep cleansing properties are equally potent detoxifiers of the human body. However, their ability to clean the oceans means they are often themselves contaminated, so care must be taken in sourcing algae from clean waters before using them as food in detoxification and purification regimes.

Thalassotherapy is one of the regimes offered by many spas. It is an ancient Greek therapy, based on a combination of sea air, protein-rich algae and mineral-rich muds and clays from the seabed. It was originally recommended as a remedy for boosting the constitution, and to combat muscular stiffness and digestive complaints. Therapies that rely on sea muds and clays are sometimes known as pelotherapy.

An increasing number of spas offer cures based on seawater, algae and mud. Treatments usually involve bathing and exercising in swimming pools filled with comfortably heated seawater. To stimulate blood circulation and lymphatic drainage, clients may be massaged under jets of hot seawater to which algae has been added, and some centres offer full body massages with a therapist working under a continuous shower of seawater. These treatments are particularly recommended for rheumatic problems.

## Seawater and Radiation

Radiation means different things to all of us but it is a natural phenomenon, part of life: we experience its benefits as light and warmth from the sun. There are potential problems with increased radiation around us in the form of electromagnetic energy from domestic and industrial sources, but in moderation natural background radiation is vital to health – it enhances cell and tissue regeneration, and generally brings about a feeling of wellbeing.

Excessive radiation is a common cause of irritability, fatigue, depression and illness; it deprives the body of sufficient red blood cells, weakening the immune system and tissue repair mechanisms. Sea salts themselves contain minute quantities of radioactive elements, and seem to have considerable ability to neutralize and balance radiation. This means that regular sea bathing, thalassotherapy or healing baths at home can help to balance our physical and emotional systems.

Salt has always been recognized as a powerful cleanser, and as a preserver of energy (for example, in keeping food fresh). Water from the Dead Sea in Israel has a greater concentration of minerals than that from anywhere else in the world – 1 litre (2 pints) of Dead Sea water contains 30 per cent salt, and an astonishing 1.1 kg (2.5 pounds) of dissolved mixed minerals.

Bathing in water with a high salt content is particularly recommended for skin complaints, for muscular and skeletal problems, and as a general constitutional cure. It is very beneficial for people recovering from surgery as the increased buoyancy allows movement that would otherwise be painful or impossible.

Dead Sea water is therefore highly magnetically charged, as well as charged through healing UV radiation from the sun. Its magnetic and electrical energies are in equilibrium, and can therefore help balance the energies of the body. The potency of its waters make the Dead Sea, the lowest point on the Earth, the most healing sea of all.

- Add two teaspoons of Dead Sea salts to a basin of warm water to bathe your skin to combat skin eruptions – acne, psoriasis and eczema.

# Healing Baths

---

### CLEANSING YOUR WATER

The easiest way to cleanse and so re-energize your bath water is by adding Dead Sea salt. You can then imprint this with healing energy through sound, scent and colour.

You need a glass container with a lid (or a jar made from natural minerals such as onyx or alabaster); two handfuls of Dead Sea salt crystals – always use natural grey salts rather than refined white crystals; and a teaspoon of unperfumed talcum powder.

- Fill your bath with warm water. Then fill your container with water from the bath, add the salts and talcum powder, put on the lid and shake vigorously for a few minutes. This mimics the succussion used in homeopathy.
- Empty the container into the bath water, stirring with your hand as you add it: make several large figure of eight movements clockwise, and then counterclockwise. This sets up a series of small vortices that help to open your bathing water to receive healing vibrations.

---

## SCENTS

A warm, fragrant bath is a wonderful way to relax, and to treat nervous tension, insomnia and muscular problems. Add a strong herbal infusion to your water, or tie a muslin bag filled with fresh or dried herbs under the tap as you run your bath. Or add a few drops of essential oils to your water, stirring vigorously to ensure thorough dispersal.

- *Lavender* is a very potent herbal bathing remedy – for relaxation, to relieve muscle fatigue, to help combat viral infections and as a detoxifier.
- *Thyme and rosemary* are good for tiredness.
- *Chamomile, frankincense and clary sage* help you to relax and sleep well.
- Never add more than 10 drops of essential oils to your bath; use only 1 drop for a baby's bath, and 2 or 3 for children

(see page 146). Chamomile and lavender are particularly soothing for babies' and children's baths. If you add them as essential oils rather than herbal infusions, first mix them into a carrier such as organic sunflower oil.
- Herbal baths are wonderful skin tonics; for soft and glowing skin add ylang-ylang, sandalwood, frankincense or any rose oil.

## THE POWER OF COLOUR

Healthy water can pick up the subtle vibrations of colour. The skin and brain are particularly sensitive to the varying energies of different colours, some of which are relaxing, others stimulating.
- As part of your healing bath experience, surround yourself with colours from the cool end of the spectrum for relaxation. Turquoise, blue or violet are particularly suitable, and green is good for balance. Let your instincts decide. You can use coloured lightbulbs, or drape the room with coloured fabrics, or choose Aura-Soma remedies.
- When you get out of your bath, wrap yourself in intensely coloured towels.
- When you have finished with your bath, be sure to clear it of your vibrations – use non-chemical cleanser, then wipe over the surface of the bath with half a lemon.

## RELAXATION

- Be sure the water is the right temperature for you – it must be warm for relaxation. The water should stay warm enough to enjoy a 15 to 20 minute bath without chilling. Do not add more hot water during your bath.

- If possible, bathe in the evening, as a preparation for a night of deeply relaxed sleep.
- Remember to close your eyes when relaxing; this allows your mental activity to slow down, and you become more receptive to the healing vibrations around you.
- Think healing thoughts. Choose to recall a joyous moment in your life, or give yourself positive affirmations for the days ahead.

---

### SOOTHING SOUNDS

Water is a powerful transmitter of sound. Our bodies can detect many frequencies that our ears cannot pick up, since the high water content of the body's tissues receives and amplifies sounds that we cannot 'hear'.

- As you relax in your bath, try listening to recorded music of sea gently rolling over sands; this will resonate with your body to promote deep relaxation. If you breathe gently and rhythmically with the sound of the waves, you will receive the healing vibrations of the sea.

# Floral Vapour Baths

The Aztecs created healing baths in special adobe or stone small rooms called *temezcals*, built next to temples. Inside the room, opposite the small entrance, was a shallow lake of hot water over heated stones, with flowers floating on the water. Herbs were thrown on the hot water and the stones to vapourize, surrounding bathers with herbal fragrances. The Aztecs used these baths as a safeguard against ill health, particularly digestive or circulation problems.

The Vedic system from ancient India contains detailed prescriptions for adding floral and herbal oils to vapour baths to combat specific complaints. They created their special floral oils by leaving bottles of oil overnight beside blossoming flowers. As a flower breathed, the oil absorbed its scent. The following night the same bottle would be left beside another flower and so on for several nights, until an exotic floral blend was created. These oils were at the heart of the Vedic system of health.

Today, we take a floral vapour bath as a luxurious and powerful treatment for fatigue, respiratory problems, treatment of infections and as a way to strengthen the constitution. Absorption of the floral oils can stimulate the oxygen in the bloodstream, hence boosting the immune system. Inhaling herbal and floral fragrances in dry heat baths also cleanses the upper respiratory system.

---

## MAKE YOUR OWN VAPOUR BATH

If you live in a hot, sunny climate you could try the ancient methods of the Aztecs or from India. Otherwise try the following:

- You need fresh scented flowers and/or herbs, a small glass container with a lid (or a small screw top jar), some natural unbleached cotton cloth and a few spoonfuls of light unrefined oil. You may be able to buy cosmetic oil from a pharmacy. The oil must not be too viscous or scented. Organic sunflower oil makes a good base, and it is full of vitamin D.
- Saturate a small piece of cloth with oil. Place it in the bottom of your container, and cover with a layer of fresh flowerheads or petals, or herbs of your choice.

- Repeat the alternate layers of cloth and flowers to fill the jar. Cover it and leave in a warm place for three or four days, agitating the jar slightly each day.
- Change the flowers, and repeat the whole process until the oiled pieces of cloth are completely saturated with the scents. The process will probably take about two weeks.
- Place your scented cloth in the bottom part of an aromatherapy vapouriser. Or you can put a steamer on a small burner, placing your cloth in the top part of the steamer. Whichever you choose, put it on a chair beside your bath, near to your head end.
- Fill your bath with about 15 cm (6 in) of warm water, adding two handfuls of Dead Sea salt crystals. Lie down in the bath and rest your head on a pillow – a folded towel or a specially designed rubber bath cushion.
- Close your eyes and relax. Imagine the tension in your body being gently released, starting with your forehead, moving down the body, and out through your fingers and toes. When you feel ready, get up slowly and dry yourself briskly in a large warmed towel.

# The Energizing Shower

Some days we awake full of energy, but on others we feel sluggish and unfocused. Occasionally, overstress, ill health or lack of quality sleep leave our energy reserves taxed to the limit. A positive way to start the day is to re-energize with a shower. The negative ions in moving water provide a feeling of energy, which is why we feel revitalized after a walk by crashing waves or waterfalls. Human-made waterfalls – showers – can help us to regain energy and mental wellbeing.

Many Northern European cultures have a long tradition of cold showers to boost circulation and general fitness as well as a positive attitude with which to greet the day. Finland, Scandinavia and many areas of Russia have highly developed water practices, combining physical benefits with a deep- rooted, but often unstated, belief in their emotional and social benefits.

The Japanese have a highly developed consciousness of water, and the spirit of cleanliness is fundamental to their religion and philosophy. Implications of ritual purification pervade the most ancient Shinto ethos and remain constant throughout Japanese history. Ritual cleansing may be as simple as rinsing one's hands at a fountain at the entrance to a Shinto shrine, or praying underneath waterfalls. Many martial arts practitioners also like to shower under a waterfall, and if there are none nearby, they may shower each morning under a bucketful of cold water.

Pine has a cleansing effect; it is a powerful antiseptic and deodorizer, and pine essential oil is recommended as a respiratory healer. But it also has other qualities. If you go for a walk in the evening among pine trees, watch them at sunset, and see how long they retain an aura of light after their neighbouring trees are in darkness. This gives a clue to the pine's ability to bring us the special energy that is important to strengthen our life force.

- Use the same mixture of Dead Sea salts and talcum powder as for your healing bath (see page 128), but this time shake them together in your container without any water. Pour this mixture onto an unbleached muslin cloth, add 4 drops of pine oil and then tie the cloth around the showerhead. As you shower, the action of the negatively charged water on your skin, combined with vibrations from the pine oil, will stimulate circulation, help you to absorb more oxygen into your system and increase your energy. Add to your shower through sound, visualization and massage.
- Begin showering under warm water, and gradually decrease the temperature, as the cooler the water, the more it refreshes and stimulates the circulation.
- Use a natural loofah or brush, gently, to further stimulate the circulation and the lymphatic system. Wet the brush thoroughly and, using circular movements, work your way over your body. Always work toward the heart for maximum effect. This opens the meridians through which *chi* energy flows, improving circulation and general health.
- Towel dry vigorously, perhaps using a brightly coloured towel from the stimulating end of the colour spectrum – yellow, orange or red.
- For a final touch, use a fine misting spray which you have previously filled with revitalized water, and add 2 drops of your favourite essential oil. Mist over your entire body and let it dry naturally.

# Healing Sound

Knowledge of healing sounds comes largely from the East, where medical practitioners believe that healthy organs resonate at the same frequency as specific sounds. When you strike a harmonic resonance at the appropriate frequency, it can improve the circulation of *chi* within the body and strengthen the constitution. These ideas have been taken up in cymatic medicine, recognizing how sound affects us by resonating with specific subtle energies in our bodies. Practitioners make use of this knowledge to direct specific healing frequencies at the body for overall health and to cure specific physical problems. So singing in the shower is more than a cliché! It really can help to refresh and invigorate you.

Chinese medical philosophies believe that there are universal sounds to express pain, illness or joy; when your voice resonates with a specific organ, it improves the flow of *chi* and strengthens the function of various organs. A system of six traditional Chinese healing sounds designates specific sounds to resonate with certain organs; the word '*hu*', for example, stimulates the spleen, the so-called 'Emperor of the body' and a central organ of our *chi* energy. Singing or chanting '*hu*' (pronounced 'whooo') in the shower may therefore stimulate your spleen, and so add to your energy.

---

## VISUALIZATION

- As you shower, imagine sunshine dappling through pine trees. Direct the colour of the sun's rays to the lower left region of the rib cage, to penetrate the spleen and draw in its vitality. In vibratory medicine, the spleen frequency resonates with the same energy as the colour and warmth of the sun.

# Massage

Chinese medicine has shown us that massaging certain points on our body can provide extra energy. Massage can be very effective in the shower, exerting pressure at certain specific points that influence the flow of energy around the body.

The thyroid gland regulates the body's biochemistry (and therefore energy), and a few moments of massage at the points connected to the thyroid gland can subtly rebalance the chemistry that sparks energy. If you do this in the shower the effect can be doubly beneficial because you are at the same time absorbing energy through the water. Two easily accessible points affecting the thyroid gland can be found at the base of the nail of each thumb, on the side nearest your forefinger. Massage from this point down an energy line towards the wishbone between the thumb and forefinger in slow small circular movements, repeating the movements several times. There are two more points at the base of the big toes (where they join the foot) that are also specific to the thyroid gland – but for safety's sake don't try and stimulate these while in the shower; wait until you are drying off.

In an Arabic *hammam*, or bathhouse, bathing is followed with vigorous body brushing, massage and mud wraps to cleanse and detoxify the system. Body brushing assists the lymphatic system, which helps to eliminate waste and toxins from the body. Unlike the circulatory system, where blood is pumped round the body by the heart, lymphatic drainage relies on movement.

# Herbs and Flowers

If you tend to suffer from morning fatigue, or are prone to catching coughs and colds, or generally lacking in energy, add to the effects of your energizing shower by massaging oils into your skin after your shower. Hyssop is particularly good to combat colds; fennel, geranium and juniper are good to detoxify the system – particularly if you have a hangover. Rosemary and bergamot are revivers, particularly useful for flagging energy; thyme and peppermint are very good to counteract any feeling of tiredness.

Spray your skin all over with rosewater after a shower. This simple beauty routine will leave your skin fresh and glowing. Once almost universally used (in kitchens as well as bathrooms) rosewater has sadly fallen out of fashion as new lotions and potions have increased in popularity. Yet it remains one of the simplest and most effective of all beauty ingredients.

- Make your own rosewater by adding fresh rose petals to a small jar of distilled water. Shake it each day for two weeks, then strain the petals and use the water. Spray it on your skin whenever you need a lift, and use it routinely after bathing or showering.

# Taking the Waters

The word spa is often used to describe a health centre that offers specific beauty treatments and exercise routines loosely based on some kind of hydrotherapy. But a spa is, literally, a place with mineral springs, named after the town of Spa in Belgium. This was one of the first places to recognize the healing potential of its mineral-rich waters. Thousands of stories attest to water from specific spas being ascribed with miraculous properties to overcome a wide range of ailments, from influenza to blindness – an Austrian tourist board brochure recently claimed their spas could treat problems ranging from developmental disorders in children to infertility problems in older women!

Every well, spring or spa is unique. Their waters vary according to the mineral and trace element contents. Such variation depends on the rocks, clays and peats that the water filters through, the natural gases released into the water, and the influences of local electric and magnetic fields.

'Taking the waters' has always been an accepted cure. In the last century in Northern Europe it was an essential part of the social calendar, but its popularity declined with the advent of improved domestic water supplies. Now spa treatments are becoming popular again, particularly among young people.

If there is no spa near you, or you want to try some of the treatments offered abroad, many travel agents offer spa holidays. As well as European destinations or the United States, spa seekers can travel to Finland, Costa Rica, Jamaica, Israel, South Africa, Thailand and Malta, as well as to some of the oldest spas in Turkey, Iceland and Japan.

Spa water is particularly beneficial today because of the mineral deficiencies brought on by unhealthy diets and lifestyles. When you immerse yourself in mineral water your body receives necessary minerals to balance health problems, and experiences a rise in the anti-inflammatory hormones, or endorphins, that relieve pain. Spa treatments usually combine the positive effects of breathing in negative ions with an intake of trace elements and minerals, often with the additional benefits of marine salts and antibacterial plankton.

If you go to a spa suffering from stress, or in pain, you are likely to leave in a state of blissful relaxation and a feeling of total wellbeing. Recommended for general good health and against stress, spa treatments are also highly beneficial for:

- *Asthma:* negative ions in the air boost the respiratory system, while minerals help regulate the production of histamine.
- *Muscle and joint pain:* increased levels of endorphins lead to reduced pain and increased mobility.
- *Circulatory problems and varicose veins:* minerals absorbed into the body boost and balance the blood stream and ease blockages in circulation.
- *Psoriasis and eczema:* antibacterial plankton and minerals balance the needs of the skin.
- *Rheumatoid arthritis:* mineralized and energetic water helps to prevent joint problems and stiffness.
- *Osteoarthritis and osteoporosis:* minerals in the water boost the immune system and stimulate growth of white blood cells.
- *Following surgery:* the antibacterial and energizing effects of spa water helps healing and recovery.
- *Minor infections:* can be cured through balancing minerals and antibacterial plankton in the water.

# Hot Springs and Whirlpools

All volcanic islands are blessed with numerous hot mineral springs, bubbling up through the earth with a turbulence that maximizes the healing energies of the water. Immersion in this hot spring water must be the most sensual of all bathing experiences: bathing out of doors in constantly bubbling water that stays hot even if the air temperature is freezing and snow lies on the ground. Bathers slip peacefully into a sense of gentle harmony with their surroundings. At popular springs people may arrive in party mood, perhaps with a group of friends or colleagues. But before long they grow quieter and slump deeper and deeper into the hot water, occasionally sighing with pleasure!

> Some spas offer treatments using blue-green algae, spirulina and chlorella. These algae contain vitamin B12, phytonutrients, antioxidants and chlorophyll that have detoxing and immune boosting properties. The blue pigment of these algae can help the body get rid of heavy metal pollutants. You can buy blue-green algae as a bath, beauty or food supplement. Their nutritious qualities are well recognized – NASA studied the possibility of using spirulina (once eaten by the Aztecs) as a space food.

Different hot springs have different mineral contents. In Japan the hot spring culture is so developed that the mineral content of every one of their several thousand springs, or *onsens*, is published and available at tourist offices. Visits to their hot springs have been popular among the Japanese throughout their history. The different hot springs are known to be effective in curing specific ailments. The Japanese visit the springs for relaxation, as an aid to overall wellbeing or specifically for healing purposes. They often bathe publicly in a group – even at home, bathing is often a family affair. Their habit of daily bathing – first washing away body dirt then soaking in a deep tub or pool of clean water – has not changed in centuries.

The development of hot tubs is an attempt to imitate the hot springs experience in a non-spa setting. First enjoyed in California, they are becoming increasingly popular. One of the central elements to a hot tub is its sociable nature. A group of people bathe together in a tub which incorporates hot jets of water – these imitate the effect of naturally bubbling waters.

Jacuzzis or whirlpool baths have become popular fittings in modern bathrooms in the US and Europe. Their outdoor equivalent, the spa bath, is popular throughout Australia where the climate suits outdoor bathing all year round. In a Jacuzzi, jets of water erupt into the water from strategically placed nozzles, so the water is always moving and picking up energy, forming little whirlpools and vortices as it meets resistance. The moving water relaxes and energizes your body; you can choose a constant temperature to suit you, or else change the temperature during your bath.

Jacuzzis are excellent as a general aid to wellbeing and health. Their massaging effect is particularly good for people suffering from circulatory problems or muscular stiffness, especially after sport as they effectively prevent the build-up of lactic acid in the body. Hot whirlpool baths help to relieve chronic pain.

The hot springs of Iceland are as popular as those of Japan. The country boasts many mineral-rich lakes – one favourite is the 'Blue Lagoon', where the water picks up colour from the minerals and muds it contains, and is believed to have numerous curative properties.

Some hot springs and spa resorts offer mud baths, where clients wallow in hot mineral-rich mud and water. Suspended and soothed between layers of mud and hot mineral water, your body will sweat and be cleansed of toxins. The heat and weight of the mud relaxes muscles and increases circulation, loosens joints and slows down the nervous system. Mud baths are also good for treating acne and psoriasis. For maximum effect, follow a mud bath with a shower and cooling bath, then rest.

Mud wraps are excellent detoxifiers: you are cocooned in a mixture of algae and seawater, coated in clay, followed by gentle scrubbing and massage. The protein- and vitamin-rich algae and muds pull toxins from the body while at the same time nourishing it. Swimming and massage finish off the deep cleansing and restore the circulation and energy.

If you can't get to a hot spring, hot tub or Jacuzzi, find a centre near you that offers hydrojet massage. Here, jets of hot water are directed all over your body as you lie on a massage mattress or sit in a specially designed bath. This is a perfect therapy for stress and tiredness, relieves aches and pains, and generates a wonderful feeling of wellbeing.

# Sweating it Out

A sauna or steam bath is a great revitalizer. Although sweating is a natural cure, avoid steam baths, saunas and very hot baths if you have high blood pressure, angina, heart disease or diabetes. Don't take any sweat bath if you have recently had an operation, suffer from epilepsy, are asthmatic or have a history of thrombosis.

Dry heat is good for respiratory conditions and as a gentle constitutional remedy. Damp heat acts quickest to discharge blocked energies associated with colds and sinus problems. Hot steam treatments are excellent for treating viral or bacterial infections, and for regulating slight fevers. Hot steam increases skin action to cause sweating, which cleanses the body from within, releasing unwanted waste through the skin's two to four million sweat glands. About 98 per cent of sweat is water, but the rest is toxins – salt, heavy metals, nicotine (if you smoke) and chemicals from the environment. Particularly good for athletes, steam also flushes out lactic acid, a major cause of stiff muscles and fatigue.

## Saunas

The modern sauna reflects ancient practices of native tribes and cultures that used to dip alternately in hot springs and icy cold pools and rivers. Steam baths have been regular features of life for centuries, if not millennia, throughout North and South America, Russia and Finland. Most steam treatments are followed by a dip in a cold pool, or a cold shower, or even a roll in the snow – the complete contrast stimulates circulation and energy, and leads to a euphoric feeling.

> Caution: Saunas and steam baths are not suitable for women in the first three months of pregnancy or the very young. Elderly people should only take saunas or steam baths if they are used to the practice.

The Finnish sauna ideally combines all elements of dry heat, wet steam, a plunge into a cool pool and, if possible, massage. The sauna is a room usually made of pine with a small fire of burning coals onto which water is ladled periodically to increase the temperature and increase the steam. Before entering the sauna, Finns like to spend ten minutes or more in a dry heat bath to start the process of perspiration. They then move to the sauna, taking with them a small pail of water. With a birch whisk they flick water onto the coals, and also use it to beat the body gently to increase circulation. This brings the blood to the surface and makes the sweat flow freely, encouraging thorough detoxification through the skin. After 30 minutes in the sauna, they like to plunge straight into a cold pool. This is a very heady experience after the initial shock – yet Finns like to return to the hot chamber, sometimes after a massage, and repeat the experience.

## Sacred Sweat

The sweat lodge has always played a sacred role in the ritual of many North American Indian tribes. Members of a community gather in a specially constructed small building, or 'lodge', to sweat. In the Native American tradition there is an intimate and deep connection with water, not just for nurturing the body, but also the soul. For this reason, sweat lodges are enjoying something of a renaissance among New Age cultures.

Taking part in a sweat lodge is an important ritual whose purpose of cleansing and purification allows a new sense of self to emerge. Cleansing is combined with profound reflection and introspection, as well as a deep connection both with the others involved in the ceremony and with the spirits of nature.

For extra healing effect, sprinkle a few drops of one of the below essential oils into the water that you flick onto the coals of your sauna:

- *Pine oil or essence* has antiseptic qualities. Pine is a herbal diuretic, and a respiratory healer and cleanser. Because it stimulates adrenaline production, it is particularly good for tiredness.
- *Lavender oil* calms the nerves and helps heal skin problems.
- *Eucalyptus oil* has antiseptic and antibiotic qualities.
- *Ginger oil* stimulates the circulation.
- *Peppermint oil* stimulates digestion.
- *Rosemary oil* combats muscular aches, rheumatism and arthritis.

## Turkish Baths

Turkish baths, or hammams, are dry heat baths where perspiration is induced by hot air from heated stones, usually also followed with bathing or showering and massage. They originated in the sensuous practices of early Arabic cultures, emphasizing the cleanliness demanded by Islam but also the belief in the body as the temple of the soul.

The original *hammams*, popular throughout the Byzantine world from *c.*500 AD, were retreats linking spiritual and physical purification. They combined numerous different chambers and halls with special treatment rooms for massages, mud wrapping and body brushing. Women could spend the whole day bathing in their *hammam*, relaxing, massaging, anointing, enjoying each other's company and gossiping.

*Hammams* still exist, some recreated in their original splendour and complexity. A visit is a sensual and luxurious experience, rather like a trip to an exclusive health centre, with hot and cold rooms, hot and cold pools, massage, aromatherapy and mud treatments.

# Floating

Floating is one bathing therapy that is becoming increasingly popular in Western society, particularly as an antidote to busy stressful lives. Most people can benefit, but don't float if you suffer from epilepsy, schizophrenia, kidney conditions, acute skin disorders, thrush or fungal infections.

Flotation tanks first became popular in Australia and the US, finding their way to Europe more recently. A bather floats in an enclosed tank or pool, without light, in water up to 25cm (10in) deep, made buoyant by the addition of about 300kg (660lbs) of Epsom salts. Epsom salts are strong detoxifiers, working through the kidneys, so be particularly careful to rehydrate yourself. Make sure you are warm and comfortable before resuming your daily routine – your body will initially be slightly weakened and you are at risk of catching a chill until your body rebalances.

Floating reduces blood pressure and heart rate. It slows your brain waves, and balances the activity of the left- and right- hand sides of your brain, leading to clearer and more creative thought. Floating in salts lowers the level of stress-related chemicals in the body and stimulates the production of endorphins. It also rids people of general aches and pains, and is increasingly finding success with sufferers of diseases such as multiple sclerosis, arthritis and ME. One centre in London is experimenting with floating as an accelerated learning tool.

Before floating some people may worry that the experience will be claustrophobic, or unpleasant, but such worries usually disappear quickly. Most people (re)discover the ability to retrieve an inner sense of harmony. Typically, soothing music is played for the first ten minutes, when most people will drift off to sleep.

A float usually lasts for about an hour, and floaters completely turn off from the world, losing touch with their day-to-day reality and worries, therefore able to access a deeper consciousness. It has been suggested that floating brings back the sense of being in your mother's womb, which is a reason why it soothes stress and tension. At body temperature, water, the universal solvent, seems able to dissolve emotional boundaries and blockages.

---

## FLOATING

- Don't fit a float into a busy schedule but make sure you allow yourself at least two hours away from your normal routine to enjoy the full benefits.
- Don't shave or wax on the day of your float – it could sting.
- Eat only a light meal before floating, as digestive discomfort could interfere with your float experience.
- Avoid stimulants such as alcohol and caffeine the day before floating.
- Cover any cuts or scratches with Vaseline before you float, and remove contact lenses.
- When you get into your tank, lie back gently, try not to splash about too much as the saline water takes time to settle.

- Don't expect to relax fully straight away.
- You don't have to lie in complete darkness; you can float with a light on if you prefer.
- Get comfortable in the centre of the tank so you don't worry about banging into the sides.
- Focus on de-stressing by deep breathing and visualizing somewhere you associate with relaxation.
- Have a good stretch before you get out to realign your vertebrae if you have been floating in a curled up position – many people do.
- Allow recovery time.
- You may feel emotional, euphoric or disorientated. Take time out to rebalance.
- Drink lots of water to rehydrate.

# Shape Up with Water Exercise

For those of us who live near a public swimming pool, water provides the perfect medium for exercise. Don't be put off because the water in all public pools has to be disinfected in some way, usually with chlorine or other chemicals. Just be aware of it. Before bathing, protect your skin with a good natural moisturizer so you will absorb less chlorine from the water. Always shower straight afterwards. Make a sachet of Dead Sea salts and talcum powder to hang under the showerhead – to cleanse and energize the shower water, and yourself.

Water exercise allows for movement that may be too painful or stressful on land. Water's buoyancy reduces body weight by 90 per cent, so stress on body parts is minimal. Water exercises are good for people recovering from injury, or who are obese, arthritic, with musculoskeletal problems, multiple sclerosis or osteoporosis. Pregnant women can exercise easily in water without putting pressure on themselves or their baby. The buoyancy of water gives your body extra support, and protects your joints from injury. The pressure of water provides added resistance and makes your muscles work harder than when you're exercising at home or in the gym.

Start any form of exercise gently with a five-minute warm up; first try a few simple kicks, head turns and shoulder rolls – just to loosen up. Water exercise can be particularly useful for people who suffer from arthritic problems or stiffness, or for elderly people who may find other exercise difficult or painful. It can also help alleviate stress, and help reduce weight.

## Watsu

Water is the ideal medium for freeing the body. Developed in California, watsu is a water-based form of shiatsu, an Eastern style of bodywork therapy that is aimed at balancing the body's energy meridians. Watsu involves gentle cradling and body manipulation in water, as well as exercises. It is usually performed with a partner, and is a wonderful way of alleviating stress or specific physical symptoms.

Eastern healing concentrates on maintaining *chi* energy in an optimal condition, or else restoring it. Manual manipulation is one way such healing uncovers the resources we have within ourselves that contribute to our state of wellbeing. Manipulation in water not only makes the body weightless and therefore more flexible than when lying on a mat; the water's energy also seems to have a positive effect in unblocking the energy channels.

Many spas ensure their pools are bacteria free by adding ozone ($O_3$) to the water rather than chemicals. This oxygenates the water to assist clearance of any toxins. Some others add diluted hydrogen peroxide to the water as a powerful natural sterilizer.

## CONSTITUTIONAL SUPPORT

- Before getting into the pool, stretch and yawn. Breathe in as you raise your arms, then circle them round. Feel the energy flow through your body as you exhale.
- Lie on your back in the water, with a partner gently supporting you with one arm. Your partner should pick up your arm, lift it above your head, and bring it round and down to your side. They should then gently press down the inside of your arm with the palm of their hand, moving gently from the top of the arm to inside the wrist.
- Your partner should then move to your other side, gently supporting you with their other arm, and repeat the process.
- Then turn your attention to your legs. Your partner should first pick up one leg, bending it toward your chest and rotating it before letting it lie back in a relaxed position and gently pressing down the inside of the leg with the palm of their hand.
- When your partner has worked on both of your legs, relax and float for a few minutes before performing the exercises on your partner.

## EASY WALKING

- Walking through water provides good all over conditioning for your legs, as you work against the force of the water. It is good exercise for stomach muscles. If you push your arms forward and round as you walk, this exercises your upper back muscles and the pectoral muscles in your chest.

## DEEP WATER RUNNING

- Make running or cycling movements under water using your hands to keep you afloat – you may need a buoyancy aid. This works all the muscles in your legs, and your arms if you use them. It helps reduce fat, and is a good way of encouraging the heart and lungs to work properly.

## LEG LIFTS

- Hold on to the sides of the pool, and lift and lower your legs in a bent position. This will contract the muscles at the front of your hip on the way up, and the buttock muscles on the way down.

# Bathing for Common Ailments

Bathing can be highly therapeutic and a great way of boosting your ability to combat common ailments. But if you are in any doubt about your medical condition, or the suitability of a specific treatment, consult your medical practitioner first.

## POINTS TO REMEMBER

- *Saline water* is a good general tonic; particularly effective for skin problems; an antiseptic and wound healer.
- *Warm water* is sedating and relaxing.
- *Steam* increases skin action and causes perspiration; can ease chest constriction and help relieve respiratory problems.
- *Whirlpool baths* both relax and energize; relieve stiffness and muscle pain.
- *Showers* restore energy and strengthen the constitution.
- *Cold water* is restorative, energizing and helps build resistance to disease. It can reduce fever, relieve thirst, aid elimination of toxins and act as stimulant, diuretic and anaesthetic. It will restrict blood flow in injuries. Cold water stimulates a sluggish immune system, and helps general constitutional health. It makes surface blood vessels constrict, restricting blood flow and sending blood toward the internal organs. Restricted blood flow inhibits the biochemical reactions that cause inflammation. As cold water decreases the temperature of a particular area, it slows the metabolic process and lessens the cells' need for oxygen and nutrients.
- *Hot water* is relaxing, and can encourage sweating to clean the body from inside. Hot water dilates blood vessels, reducing blood pressure and increasing blood flow to the skin and muscles, improving circulation, and easing stiffness. Improved circulation can also boost the immune system, sending more oxygen and nutrients into the tissues to repair damage, and helping to remove waste products from the body.
- *Hot and cold baths* stimulate circulation and speed up body responses. Bathing alternately in hot and cold water stimulates the hormonal system, reduces circulatory congestion and balances blood pressure by narrowing and dilating

the blood vessels. It can relieve inflammation by relaxing and contracting the affected part. It also increases the body's electrical potential – and therefore its ability to receive vibrations or energy – and it stimulates the lymphatic system into activity which encourages efficient waste-clearance.
- Caution: Never apply heat to any acute injury – heat increases the blood flow and metabolic rate and therefore increases the inflammatory response and production of tissue fluid.

## Compresses

Compresses are often used alongside bathing to relieve pain, restore mental and physical energy, and heal wounds.

- *Cold compresses* stop bleeding, relieve pain after injury, relieve joint pain, reduce head congestion and check inflammation. Cold compresses can be single (cooling), or double (heating), where a dry cloth is placed on top of the cold compress, sealing in the moisture that gradually heats up with the body temperature, keeping body heat in.
- *Ice compresses* are particularly suitable for wounds, sprains and strains – never put ice straight on to the body; always wrap it in a cloth, or use a bag of frozen vegetables!
- *Hot compresses* can relieve pain, stimulate perspiration, improve circulation of a local area and help rheumatic complaints.

## Common ailments

- *Arthritis* Steam bath; cold double compress; spa treatments; sauna; mud baths or mud packs; flotation.
- *Asthma* Spa treatments; body brushing.
- *Backache* Spa treatments; flotation; mud baths.

- *Bites and stings* Cold compress with essential oil of melissa (lemon balm).
- *Bronchitis* Steam baths.
- *Burns* Cold water followed by cold compress. As soon as the burn shows the first signs of healing, apply several drops of wheat germ oil – its high vitamin E content helps prevent scarring.
- *Circulation* Spa treatments; saunas are particularly effective when combined with cold showers; whirlpool baths; alternating hot and cold baths or showers. Add ginger to bath water to stimulate circulation further. Warm foot bath – cover the bowl around your feet with a towel to keep in the heat: add three teaspoons of mustard powder for chilblains and poor circulation.
- *Colds* Sauna – damp heat helps discharge blocked energies associated with colds and sinus trouble.
- *Constipation* Shallow cold bath – add a teaspoon of vitamin C to overcome genital or bladder infections, or add a few drops of eucalyptus, juniper or chamomile oil.
- *Cystitis* Hot shallow bath or neutral bath with a cupful of cider vinegar or a few drops of tea tree essential oil.
- *Depression* Spa treatments; energizing shower; cold foot baths; cold compress to head; warm bath with oil of basil.
- *Detox* Hot bath with a cupful of epsom salts; spa treatments, especially marine healing with algae and muds; flotation.
- *Diarrhoea* Cold baths with hayflower or cider vinegar added.
- *Eczema* Sea bathing; bathing in water with dead sea salts added; spa treatments; saunas.
- *Eyestrain* Bathe eyes with warm decoctions of eyebright or echinacea for tiredness; add chamomile, golden seal, or yarrow if you have an allergy problem, or use chamomile teabags!
- *Fatigue* Cold double compress to feet with cider vinegar or favourite herbs added; cold float baths; sauna – add pine oil

to the water that is poured over the coals as it stimulates adrenaline production; cold bath; sea bathing; floral steam bath; flotation.

- *Haemorrhoids* Alternate hot and cold baths.
- *Headaches* Cold compress on the back of the neck for a tension headache; cold compress on the forehead for a thumping headache – add a drop or two of chamomile essential oil; warm foot bath – cover the bowl around your feet with a towel to keep in the heat and add 2–3 drops of pine oil for sinus headaches, or peppermint for digestive headaches.
- *Insomnia* Warm bath before bed, with a few drops of oil of myrrh if your mind is overactive; or add chamomile, valerian or passiflora.
- *Menstrual pain* Hot shallow bath.
- *Muscle spasms, stiffness, aches and pains* Warm bath; warm compresses on affected parts; whirlpool baths; spa treatments; mud baths or mud compresses; flotation.
- *Respiratory problems* Turkish baths or floral steam baths – dry heat is particularly effective for respiratory conditions and release of the body's impurities; sauna with pine oil added to the water that is poured over the coals – pine clears respiratory channels.
- *Rheumatism* Whirlpool baths; spa treatments; mud baths or mud wraps; flotation; hot compress (or hot water bottle) to affected parts.
- *Sciatica* Hot compress followed by short cold shower; mud baths; floral steam bath.
- *Skin irritation* Warm oatmeal baths – place several handfuls of uncooked oatmeal in a muslin bag, hang it over the tap so the water runs through it, then use the bag gently as a sponge; cold compress over the affected area cools the blood and helps to calm inflamed skin – 1–2 drops of geranium oil will be soothing.

- *Skin problems* Sea bathing; add dead sea salts to your regular bath; facial saunas or steam baths – for blackheads and blocked pores add a few drops of patchouli oil, which has excellent antiseptic and fungicidal qualities; spa treatment.
- *Sprains and strains* Ice compresses; alternate hot and cold compresses with arnica tincture to minimize healing time: always use cold compress first.
- *Stomach ache* Hot shallow baths act quickly to alleviate pain in pelvic area or abdomen.
- *Stress, tension, anxiety, worry* Neutral (body temperature) or slightly warm bath with a few drops of rose oil, then lower temperature to cold; warm showers sedate the central nervous system; hot moist compress against the spine; steam bath for the hands, with a few drops of essential oil of myrrh; colour baths or Aura-Soma; flotation; cold compress to the head eliminates mental exhaustion and helps relieve depression; a chosen Bach Original Flower Remedy can be added to any bath for gentle emotional healing.
- *Toothache* Alternate hot and cold foot bath.
- *Vaginal infections* Warm bath with cider vinegar.
- *Wounds* First bathe with cold salt water to disinfect, then bathe regularly with calendula tincture to aid healing; an ice compress decreases the amount of pressure in the capillary vessels and lowers the extent of the bleeding into the tissue spaces. Never treat a wound with hot water – this will further encourage the bleeding.

# Children's Ailments

- *Colic* Warm compress to the abdomen with dilute oils of chamomile or lavender, lemon balm or fennel to help relax the bowel; warm compresses to the chest.
- *Croup* Steam baths with added oils or infusions of either pine, eucalyptus, lavender, chamomile or catnip; hot herbal bath before bedtime.
- *Fevers* Bathe in tepid water with infusions of yarrow, limeflower, elderflower or chamomile – add echinacea to enhance immunity; (see also Kneipp, page 115-117).
- *Sleeplessness* Warm herbal bath with dilute oils of lavender or chamomile, or strong infusions of either limeflower, catnip or lemon balm.
- *Whooping cough* Hand and foot baths with strong infusions of either elecampane, coltsfoot, thyme in early stages; later add thyme, cypress, eucalyptus, lavender or marjoram.

# First Aid

- *Bruises* Cold compress with distilled witch hazel, pot marigold or arnica.
- *Minor cuts and wounds* Regular bathing in cold water with golden seal, calendula or myrrh added.
- *Nose bleeds* Cold compress of distilled witch hazel or yarrow; cold hand bath.
- *Sore throat* Cold double compress.
- *Sunburn* Cold compress over damaged area. As temperature of the skin begins to cool, apply aloe vera gel.
- *Swelling* Cold compress with hayflower.

# Aura-Soma

Aura-Soma is a system of colour therapy using colours created from the living energy of plants, minerals and light. Liquid colour combinations are made by combining an oily part of a plant with a watery part, to which are added the energies of crystals and gems from the mineral kingdom, and the energies of colour and light.

Ninety-six colour combinations are recommended for different stages of personal and physical and emotional growth, and the bottles of liquid colour can help individuals balance their emotional and physical needs.

Choose your colour combination, and shake the bottle for a few seconds to mingle the layers of oil and water before adding a few drops to your bath. As the two layers mingle, there is a balancing of energies from the plant and mineral worlds, and this balance is then carried in the water as part of your bath experience. Aura-Soma colours are created from the plant and mineral world, they are magnetically and electrically balanced, and can act as a very direct and effective way of balancing emotions and energies.

# Also available in the series:

**The Healing Nature of TREES**
PATRICE BOUCHARDON

**The Healing Nature of EARTH**
LIZ SIMPSON